ROVER SD1

Other Titles in the Crowood AutoClassics Series

AC Cobra	Brian Laban
Alfa Romeo Spider	John Tipler
Aston Martin: DB4, DB5 and DB6	Jonathan Wood
Aston Martin and Lagonda V-Engined Cars	David G Styles
Austin-Healey 100 and 3000 Series	Graham Robson
BMW M-Series	Alan Henry
Datsun Z Series	David G Styles
Ford Capri	Mike Taylor
The Big Jaguars: 3½-Litre to 420G	Graham Robson
Jaguar E-Type	Jonathan Wood
Jaguar XJ Series	Graham Robson
Jaguar XJ-S	Graham Robson
Jaguar XK Series	Jeremy Boyce
Jaguar Mk1 and 2	James Taylor
Jaguar S-Type and 420	James Taylor
Lamborghini Countach	Peter Dron
Land Rover	John Tipler
Lotus and Caterham Seven	John Tipler
Lotus Elan	Mike Taylor
Lotus Esprit	Jeremy Walton
Mercedes SL Series	Brian Laban
MGA	David G Styles
MGB	Brian Laban
Morgan: The Cars and The Factory	John Tipler
Porsche 356	David G Styles
Porsche 911	David Vivian
Porsche 924/928/944/968	David Vivian
Range Rover	James Taylor and Nick Dimbleby
Sprites and Midgets	Anders Ditlev Clausager
Sunbeam Alpine and Tiger	Graham Robson
Triumph TRs	Graham Robson
Triumph 2000 and 2.5PI	Graham Robson
TVR	John Tipler
VW Beetle	Robert Davies
VW Golf	James Ruppert

Rover SD1
The Complete Story

Karen Pender

First published in 1998 by
The Crowood Press Ltd
Ramsbury, Marlborough
Wiltshire SN8 2HR

© Karen Pender 1998

All rights reserved. No part of this publication may be reproduced or transmitted in any form or by any means, electronic or mechanical, including photocopy, recording, or any information storage and retrieval system, without permission in writing from the publishers.

British Library Cataloguing-in-Publication Data
A catalogue record for this book is available from the British Library.

ISBN 1 86126 111 X

Picture Credits

Photographs supplied by Mirco de Cet and the author.

Typeface used: New Century Schoolbook.

Edited and designed by
D & N Publishing
Membury Business Park, Lambourn Woodlands
Hungerford, Berkshire.

Printed and bound by The Bath Press

Contents

Acknowledgements	6
Introduction	7
1 Ancestors – Rover, Triumph and Standard	9
2 The Rover V8 Engine	17
3 The Birth – From P10 to SD1	22
4 The New Rover 3500	43
5 The Six-cylinder SD1 Cars – 2300 and 2600	59
6 Wheel of Misfortune – Making the SD1	71
7 The Rover V8-S – the Luxurious SD1	83
8 Rover 2300, 2600 and 3500 Models – 1980–1986	89
9 Rover 2000 – a Famous Name Revived	101
10 The 2400SD Model – the Rare Turbodiesel SD1	107
11 Rover Vitesse – the Ultimate SD1?	113
12 The Rover SD1 in Motorsport	127
13 The Police Cars	141
14 Overseas Variations and Other Unusual SD1 Cars	153
15 The End of the Line	165
16 Rover SD1 – the Poor Man's Ferrari?	171
17 Living with the Rover SD1	176
Appendix I: SD1 Chassis and Engine Numbers	183
Appendix II: SD1 Colours and Trim	187
Further Reading	189
Index	190

Acknowledgements

It has taken me several years to carry out the necessary research for this book. I would like to thank all the people I have talked to who have kindly provided me with the information I needed in order to be able to tell the complete story of the Rover SD1.

In particular, I would like to thank Graham Robson for helping me to get the project started in the first place and Mike Lewis, whose record-keeping helped to make my task much easier.

I would like to thank all the people I have talked to – past and present Rover employees, most of whom were closely involved with the SD1 project. These include: Roy Axe, John Bacchus, Dennis Barbet, Denis Chick, Mike Cook, Alan Edis, David Eley, Nigel Heslop, Spen King, Graham Lewis, Mike Loasby, Phil Mander, Rex Marvin, David Nicholas, Geoff Purkis and Ken Stansbury. I would also like to thank the people I have talked to who were concerned with the SD1 in motorsport: Ian Beveridge, John Davenport and Tony Pond.

Thanks also go to a number of classic car clubs: Dan Worley and the Rover SD1 Club, the Rover Sports Register, the Standard Motor Club, Club Triumph and the Pre-1940 Triumph Owners Club.

I would also like to thank Ian Hyne, Richard French and the Metropolitan Police, the West Yorkshire Police, Coleman Milne, Janspeed, the Department of Transport, the RAC Motor Sports Association, Rover Group/BMIHT, the National Motor Museum and the Ford Motor Company.

I am also grateful to Anders Clausager and Karam Ram at the British Motor Industry Heritage Trust archives, and to Annice Collett at the National Motor Museum for their help while I was working on this project. I would also like to thank Mirco de Cet for taking the photographs.

Introduction

A few years ago, my husband Richard and I owned a small, family saloon (a Skoda in fact – no jokes, please!). We decided that we needed a larger car, with good luggage space, and which would be comfortable to travel in during long journeys. Having looked at most of the alternatives – Mercedes (too expensive), BMW (appeared to be over-rated), Ford Granada (not terribly quiet or impressive) – we settled for a 1984 model Rover 3500 Vanden Plas. This was the start of our love/hate relationship with the Rover SD1, which continues to this date. After a few irritating breakdowns, and ultimately a noisy rear axle which needed imminent replacement (at vast expense), we sold the car. We then decided to get a more economical car, a Ford Orion, which turned out to be one of the most boring cars we had ever owned. To be fair, unlike the Rover, it was boringly reliable, although we only kept the Orion for a few months.

After this, we decided that we had to get another Rover. We eventually bought a late Vitesse, which exhibited a number of the SD1's usual problems (water leaks in the boot, electrical problems including the electric windows failing, pieces of trim falling off, and so on). It is true that the SD1 has acquired a reputation for falling apart: on the BBC's *Top Gear* television programme, Jeremy Clarkson commented that, though the SD1 was a good looking car, and went well, 'Rover didn't make them very well'! I don't know why the programme producers picked on the poor SD1 – I can think of other types of cars (certain Italian and Japanese models, for instance) which suffer badly from rust and disintegrate far quicker than the British car.

SD1, the code-name given to the Rover by British Leyland, stands for Specialist Division Number One. Most SD1 enthusiasts are aware of this fact. However, many people connected with the motor trade persist in calling the car the SD 'eye', much to my continual irritation. The British Leyland Group can be blamed for this error – the sanserif typeface chosen to depict the code-name was very ambiguous. When anyone refers to the Rover as an 'SDI', it makes me wonder whether the car was ever part of the American Strategic Defence Initiative – the so-called 'Star Wars' programme. A strange image tends to spring to mind, that of a Rover in outer space, with its bonnet aimed at Moscow. (No doubt, most SD1 owners would like to send their cars into outer space from time to time!) The Rover V8 engine, as most SD1 owners know, originated from an American Buick car – however, I think that I am safe in my belief that the Rover was never intended to form part of the Star Wars programme!

The Rover SD1 is a Leyland enigma – a mysterious, or puzzling thing which is difficult to understand. When it is behaving itself, it is a joy to drive or ride in. However, when it is not behaving, it is appalling. There are many features of the Rover SD1 which make it one of the best British cars ever designed. The car's Ferrari Daytona looks and its performance, particularly in the terrific V8-engined version, are its two main saving graces. As many owners will know, Rover SD1s can be addictive.

Introduction

In many ways, however, the Rover SD1 story is a sad one and reflects the troubles afflicting the ailing British Leyland Group at that time. However, as the last proper 'British' Rover car model before the link with Honda, I feel that a complete history of the car is now long overdue. The SD1 is much worshipped by enthusiasts, who believe that its replacement couldn't hold a candle to it. I hope that the enthusiasts will enjoy this story.

1 Ancestors – Rover, Triumph and Standard

The Rover SD1 was introduced in 1976, but its heritage can be traced back through many years of British motor industry history. The car's ancestors include not only Rover models, but also Triumphs and even Standards.

The Rover name first appeared as long ago as 1884, on a tricycle designed by John Starley. Most modern bicycles are derived from Starley's revolutionary Rover safety bicycle of 1885. In fact, the Polish word for bicycle is *rower* (pronounced 'rover').

The very first Rover car appeared in 1904 and was designed by Edmund Lewis. This car was a light two-seater with an eight horsepower engine of 1,327cc. It was unusual in having a tubular backbone frame.

The traditionally upright Rover cars of the 1930s, such as the Rover Ten, were carriages for the professional classes. They gave the Rover Company a high-class image, and were robustly made, high-quality vehicles, which sold for reasonable prices when compared with their Rolls-Royce and Bentley contemporaries.

The sporting heritage of the Rover Company goes back to the period before the First World War, when Rovers were successful in speed trials. Then came the famous 'Blue Train Rover' – the Light Six Sportsman's Coupé which raced the 'Blue Train' Continental Express from St Raphael on the French Riviera to Calais in January 1930, winning the race by twenty minutes. Thus, the SD1 was not the first sporting Rover to be built by the Company.

During the 1930s, the brothers Spencer and Maurice Wilks took charge of the Rover Company, and moved the company up-market. This was an extremely profitable period for Rover with annual production rising from 5,000 to 11,000 between 1933 and 1939. Rovers of the 1930s were well-built and had elegant bodywork. They were faster than average for the time because of their ohv engines. They certainly justified Rover's new motto – 'one of Britain's fine cars'.

The spiritual forerunners of the SD1 range were the P2 models of the 1930s (and later post-war P3 cars, which were very similar in appearance). These cars, like their later SD1 relatives, were attractive sporting saloons, fairly low to the ground, giving the driver a low, sporting driving position.

Then came the post-war P4 (or 'Auntie'). The 'Auntie' is the beloved Rover of crusty old colonel types – the sort of people who might take one look at an SD1 and announce 'Good God, Sir, call that *thing* a Rover?' The P4 was the complete opposite of the SD1. It was extremely well made and totally reliable, but looked frumpy. The SD1, as all owners know, was badly made and frequently unreliable, but had drop dead gorgeous good looks. Even now, over twenty years after its launch, most people will conclude that the Rover SD1 is a striking design.

The P4 epitomizes Rover's reputation for solid, well-engineered cars. When introduced it cost £1,106, which represented good value for money. It attracted buyers

This is a sporting Rover P2 model of the late 1930s, and, for its day, has rakish, low-slung lines.

who could not afford a Rolls-Royce or Bentley, but would not entertain the idea of buying a Jaguar (which had a reputation of being something of a spiv's car at the time). The P4 was quiet, had a comfortable wood and leather interior, and could do over 80mph. When production stopped, in 1964, over 130,000 P4s had been made.

The Rover Company had an excellent, experienced engineering team by the 1950s and went from strength to strength. Charles Spencer ('Spen') King, who became one of the key people involved in the SD1 project was with Rover by this time, having joined in 1945.

The P5 model was introduced in 1958, and to begin with came with a three-litre version of the P4 engine. This luxury car had all the usual Rover attributes including a traditionally British interior of leather and wood. In spite of being Rover's first monocoque body design, it was significantly heavier than its P4 predecessor. The other important fact about the P5 was that it was the first Rover to be styled by Rover's new stylist, David Bache, who went on to fashion the appearance of all subsequent Rovers up to the SD1. From 1967, the P5B model was fitted with Rover's version of Buick's 3.5-litre engine, (the 'B' denoting Buick). Over 69,000 P5 models had been made when production ceased in 1973.

The P5 and later P6 cars (which included the Rover 2000) were both fairly up-market. The P5B cars were used a great deal by prime ministers and cabinet ministers, up to the Thatcher era. The attractive-looking and spacious P5 is very much suited to being driven by a chauffeur. The P6 was a much smaller car physically than the P5. It was not intended as a direct replacement, but was introduced to extend Rover's range downwards.

The P6 first appeared in 1963, launched by Rover's new managing director, William Martin-Hurst, and was the SD1's immediate predecessor. The Rover 2000 offered buyers luxury and high performance in a compact saloon. It was aimed at a younger, more ambitious customer, likely to be an executive or professional person, who wanted a smart fast car and who probably would not consider buying the outdated P4. The car was an advanced design: it had a complicated suspension – a De Dion rear axle, with inde-

Ancestors – Rover, Triumph and Standard

This is a typical 'Auntie' Rover P4 model of the 1950s. Being a Rover 60, it is considered to be the Cinderella of the range by some people because it has four cylinders rather than the more usual six – however, with its lighter engine it is said to handle better.

The P5 was the first car produced by the Rover company to have a unitary body. This particular example, a P5B, was the first Rover model to be powered by the ex-GM V8 engine.

pendent coil-spring front suspension. The P6 had radial tyres and disc brakes on all wheels. It also had many safety features, including a collapsible steering column as well as interior and fascia padding. Dunlop chose it as the first car to be supplied with the Denovo run-flat tyres (also available on early SD1s). In 1965 the Rover 2000 won the AA's gold medal for its contribution to safer motoring. The P6 was eventually produced with the V8 engine option. When fitted, the resulting 3500 had a top speed of up to 122mph, and provided the rapid speed many customers had demanded (especially police forces, who bought fleets of these cars). Over 300,000 P6 models were produced in total.

The disadvantages of the P6 were the complexity of its design (the SD1 was a far simpler design), with resulting high manufacturing costs, unnecessary weight and a propensity for serious structural rust. Like the SD1, it acquired an appalling reputation for unreliability.

Before the SD1 era, the Rover Company was in every sense of the word a family company. As well as brothers Spencer and Maurice Wilks, the managing director William-Martin Hurst was Maurice's brother-in-law. Peter Wilks, who was Maurice's nephew, was also a director. Spencer Wilks's nephew Spen King was BL's director of engineering at the time of SD1, and he played an important part in the development of this car. In the words of numerous people who worked at Rover, 'Everyone knew everybody else'. This aspect of work changed altogether for the worse after the Leyland takeover of Rover. This came about through the desire of the truck-maker to gain access to Rover's knowledge in the field of automotive gas turbines, which at that time were considered to be the truck engine of the future.

A very important date in the history of the Rover Company was 1968. In that year, Rover, Triumph, Jaguar, Austin and Morris became part of the vast, disorganized and sprawling British Leyland organisation – something which had a major effect on the way the Rover SD1 turned out. Former sales manager Sir Donald Stokes became chairman of British Leyland following the Leyland/BMC merger. He was made Lord Stokes in 1969, and eventually handed over the company to a new chairman after Leyland was nationalized in 1975. However, this was not before he (along with the rest of the board) had instigated the Rover SD1 project, dictating the brief for the design team.

The Rover Company has also produced vast numbers of four-wheel drive vehicles during the post-war period, and in truth, these were largely responsible for the company's success before the Leyland era.

With regard to motorsport, Rover's racing heritage goes back to the turn of the century, when they won the Tourist Trophy in 1907. The gas-turbine powered Rover BRM appeared at Le Mans in the twenty-four hours race in 1963 and 1965 (Jackie Stewart and Graham Hill finished in tenth place in the latter race). Before the SD1, both P5 and P6 cars had been rallied with some success. However, after 1971, Rover did not take part in motorsport until 1980, when the company made a concerted effort to put the Rover marque at the forefront of British and European motor racing.

The SD1 ancestry also owes a great deal to the Triumph Company. Indeed, the SD1 has even been described ironically as 'Rover's real Triumph'! By the time of the SD1's design and development, Triumph was a part of the British Leyland Group, and many people from the Triumph part of the company were involved in the SD1 project. Within British Leyland, the merger of Rover with Triumph was announced in 1972.

From the beginning, Triumph had gained a reputation for making attractive-looking and sporty models which were aimed up-market. In the 1930s, the Vitesse name was applied to sporting Triumph models, being revived during the 1960s for the six-cylinder version of the Triumph Herald. The Vitesse name was always applied to more powerful versions of various cars – eventually including the Rover SD1.

Triumph was taken over by Standard in 1944, and whilst post-war Triumph cars were based completely on Standard Motor Company engineering, the Triumph name was retained across the range, to counter falling sales caused by the more down-market image of Standard.

Triumph is probably best-remembered as a maker of sports cars. Post-war Triumph cars had a very strong competition history, especially in rallying. TR2s competed successfully in rallies and races, including events at Le Mans. Even the humble Spitfire was rallied and raced. The Triumph Dolomite Sprint, with its sixteen-valve engine, was one of the revelations of the early 1970s, and was a great success in motor sport, especially in racing. Later, TR8 cars were rallied successfully, and competed in North American SCCA production races. Eventually, when the TR8 was withdrawn from rallying in 1980, John Davenport, who was director of motorsport at BL, needed something to replace it. The replacement for rallying and racing was not a Triumph, but the V8-engined version of the Rover SD1.

The immediate Triumph predecessors of the Rover SD1 were the Triumph 2000 range of cars – these were replaced by the six-cylinder engined Rovers. The Triumph 2000 had been a direct competitor in sales terms to the Rover 2000, being an up-market family car offering luxury and high performance. The Triumph was a more conventional design than the Rover 2000 – its smooth six-litre engine came from the last Vanguard, and it had disc brakes at the front only. However, it was cheaper than its Rover rival. In 1969, the Triumph was given a fuel-injected engine of two-and-a-half litres capacity.

There was heated argument, both at the time the cars came out and later, between people as to which car was the better of the

Uncommon when new and rare now, the Triumph Gloria Vitesse tourer of the 1930s was the first Triumph model to bear the 'Vitesse' name.

Ancestors – Rover, Triumph and Standard

These two one-time rivals are both facelifted examples of the Rover 2000 (above) and Triumph 2000 (below). The two models were first produced when neither company was connected to the other – they show two different ways of attracting the same customer.

two – the Rover 2000 or the Triumph 2000. The Triumph had its smooth, tried and tested six-cylinder engine, was slightly cheaper to buy and easier to maintain. The Rover 2000 offered more advanced engineering features and a slightly more up-market image. The disadvantages of the Rover included the fact that it was more difficult for owners to repair and maintain (the rear brake pads, for example, being a real swine to change). It all came down to personal choice.

Other Triumph relatives passed on some of their problems to the six-cylinder Rover SD1 cars. Some Triumphs had acquired something of a reputation for unreliability – particularly the Triumph Stag and the Triumph Dolomite cars, which suffered from cylinder head problems. This heritage the six-cylinder SD1 cars could well have done without.

The V8 versions of the SD1 shared the same engine, back axle and transmission as the TR8 sports car. On the six-cylinder Rover engine, the valve gear layout is a simplified version of that found on the Triumph Dolomite Sprint. The Rover six-cylinder engine is in fact derived from the Triumph 2000 six-cylinder engine, although it evolved into something completely different. The six-cylinder engine, back axle and 77mm gearbox were designed entirely by Triumph. The gearbox was also designed to be used in a range of other vehicles, including TR7s and Jaguars, and was modified for use in the four-wheel drive Range Rover.

In its body engineering, the influences on the SD1 were largely Triumph rather than Rover – it resembled the Triumph 2000, Dolomite and Spitfire models far more than its Rover P6 predecessor in this respect (for example, the way the suspension turrets and so on are arranged). The SD1 does not have any bolted-on panels – the whole body is in one piece and is welded – unlike its Rover predecessor, the P6, which had all its outer body panels bolted on to its base structure. The SD1 is a lighter car than the P6, and in this respect follows Triumph practice.

One particular version of the Rover SD1, the Standard 2000 model which was assembled in India during the 1980s, had an even stronger link with the Standard side of the post-war Standard-Triumph Company. Instead of using any of the UK-market SD1

The substantially engineered but dynamically flawed Standard Vanguard of 1947, unbelievably, was to provide the basis of the power unit in the very last variant of the SD1 to be conceived. (Photograph provided by Roger Morris.)

engines, this Indian SD1 had its own peculiar engine, the roots of which can be traced back to the Phase I Standard Vanguard of the 1940s. The Standard Company post-war had based its reputation on a one-model policy based around the Vanguard. This particular model had been intended to be 'a world car', and replaced all other Standard models in 1948. Frumpy like the P4 Rover, it did not share the refinement and finesse of that car. The Standard Vanguard's overhead valve engine was, however, extremely versatile, and powered Triumph TR sports cars, the Ferguson tractor, and eventually the Indian version of the SD1. By the 1960s, the poor sales performance of Standard cars had caused this (by now) ailing company to fall into the hands of Leyland Motors. In 1969, the 'powers that be' at British Leyland started to talk about phasing out the Rover P6 and replacing it with P10. This car was ultimately to replace the Triumph 2000 range of cars as well. The P10 project was eventually renamed SD1.

Triumph versus Rover

	Rover 2000	*Triumph 2000*
Years on sale	1963-77	1963–77
Total production	248,959	316,962
Average annual production (approx)	17,800	22,600
Construction	Base unit monocoque, with bolt-on skin panels	Conventional monocoque
Overall length (in)	178.5	173.75
Overall width (in)	66	65
Overall height (in)	54.75	56.0
Wheelbase (in)	103.4	106.0
Unladen weight (lb)	2,767	2,576
Engine	4-cylinder, overhead cam. 90bhp	6-cylinder overhead valve 90bhp
Transmission	4-speed, no overdrive. optional automatic from 1966	4-speed optional overdrive. optional automatic from 1964
Suspension	IFS, De Dion rear	IFS, IRS
Steering	Worm and roller	Rack and pinion
Wheels/tyres	6.50-14in radial-ply	6.50–13in cross-ply
Brakes	Discs at front and rear	Discs at front, drums at rear
UK Retail price (January 1964)	£1,265	£1,094

(Source: Graham Robson, *Triumph 2000 and 2.5PI*)

2 The Rover V8 Engine

The story of how Rover came to acquire the rights to manufacture a Buick V8 engine is an intriguing one.

The American motor industry has been noted for the huge number of V8-engined cars it has churned out over the years. Henry Ford brought the luxury of V8-engined cars (with their good torque) to many ordinary Americans during the 1930s, with his famous V8 models ('You'll never be late, if it's a V8' – a slogan which possibly may not always apply to the Rover SD1). From the 1950s onwards, Detroit mainly produced large saloon cars with V8 engines, usually of 5 litres or greater cubic capacity. Even quite mundane American cars, such as Chevrolets, had V8 engines. Therefore, it should be no surprise that Rover's V8 engine came from the land of the gas guzzler.

Rover had been experimenting with the designs of various five- and six-cylinder engines for some time. However, none of

The 1961 Buick Special may have been 'compact' by contemporary American standards, but it was big by anyone else's!

these experimental engines proved to be very successful, and they were never used in any production Rovers. The company was actively seeking a new power unit for its three-litre P5 model, and eventually for the P6 cars.

The new power plant was happened upon entirely by accident. Rover's managing director, William Martin-Hurst, came across it when he was visiting Mercury Marine at Wisconsin in 1963. The purpose of this visit had been to persuade Carl Keikhaefer to buy some Rover engines for marine use. There were further visits, until one day, as Martin-Hurst related (in *The Rover Story*):

> I was in his experimental workshops in Fond du Lac. in Wisconsin, talking about this and that, when I saw that lovely little light alloy V8 engine sitting on the floor. I said, 'Carl, what on earth is that?', and he told me it was for a racing boat, and that he'd originally winched it out of a Buick Skylark car. I asked him whether it would be available, and I was astounded when he told me that General Motors had just taken it out of production!

Martin-Hurst measured the unit, and found that it would fit in to both the P5 and P6 cars, being only about an inch or so longer than the Rover 2000 four-cylinder engine and slightly heavier, although, it was much wider. The only worry was, why had General Motors discontinued it? The answer was easy – there was nothing actually wrong with this engine, as Rover found out. General Motors had developed this light alloy V8 engine during the 1950s for use in a new range of compact cars. However, after producing 750,000 engines and fitting them to Buick Special, Pontiac Tempest and Oldsmobile F85 Cutlass vehicles, the engine was replace by a cheaper but larger cast-iron unit. American buyers, it appears, had begun to stop buying the smaller, so-called 'compact' cars. The old unit was too expensive for General Motors to carry on manufacturing. Production stopped in 1963.

The 1961 Buick Special was intended by General Motors to be one of their new compact models, designed to compete with the growing threat from European imports – of which the Volkswagen Beetle was by far the most successful. The Buick Special may have been compact when compared with normal American cars, but it was not exactly small by British car standards. Buick's sales department boasted of this new-sized car.

> Big car comfort and styling are combined with the economy of the small car in Buick's entirely new, lightweight Special four-door sedan, powered by the first American aluminium V8 engine in the automobile industry. The Buick Special is mounted on a 112 inch wheelbase, is 188 inches long and weighs only 2,700 pounds, some 1,600 pounds lighter than conventional-size Buicks. Its high-compression V8 engine develops 155 horsepower and gives it lively performance comparable to the bigger models in the Buick line. The Buick Special comes in two body styles, a four-door sedan and a four-door wagon, with a deluxe version offered in each. A newly designed dual-path turbine drive transmission is offered as optional equipment.

The Buick Special was priced at between $2,500 and $3,000, depending on customer specification (this was around the same price as early US Rover 2000 cars). The car was relatively economical by American car standards – drivers could get around 25mpg if they drove carefully. It was quite lively, reaching 0–60mph in under eleven seconds. The car had drum brakes and worm and sector power steering. It is likely that it had the advantages of other American cars of its era

– it would have been quiet and comfortable to travel in on long journeys. However, the Buick probably also had the disadvantages of American cars of this era – over-light steering which lacked feel and poor handling with its under-damped suspension.

For William Martin-Hurst, the old General Motors 3,528 cc unit had a number of attractions. It would fit into both of Rover's current models and was more powerful than Rover's own engines. It was lightweight and strong, and would be reasonably cheap and easy to produce.

At first, Martin-Hurst was faced with the problem that no one at General Motors seemed to take seriously his wish to acquire the rights to manufacture the engine. However, by 1964 Martin-Hurst had shipped the engine he had found at Mercury Marine back to Solihull for examination by Rover engineers. Rover were eventually given a licence to manufacture the engine (with generous terms) in January 1965. The V8 engine then appeared in production Rovers as soon as possible. Rover were fortunate to be able to acquire the services of Joe Turley, chief engine designer at Buick, who was due to retire in eighteen months' time. He was persuaded to come to Solihull as a consultant during the development of the V8 engine, to sort out any problems.

Rover made a number of modifications. Buick's Rochester carburettors were replaced by SU ones, and Lucas ignition replaced that by AC-Delco. The American automatic choke was replaced by a manual choke control. The main changes concerned production engineering – instead of die-casting the blocks with the cylinder liners held in place, Rover chose to have sand-cast blocks, with press-fit liners. Although the Anglicized version of the V8 engine used in the P5B was more powerful than the American original, it was still nowhere near being fully developed,

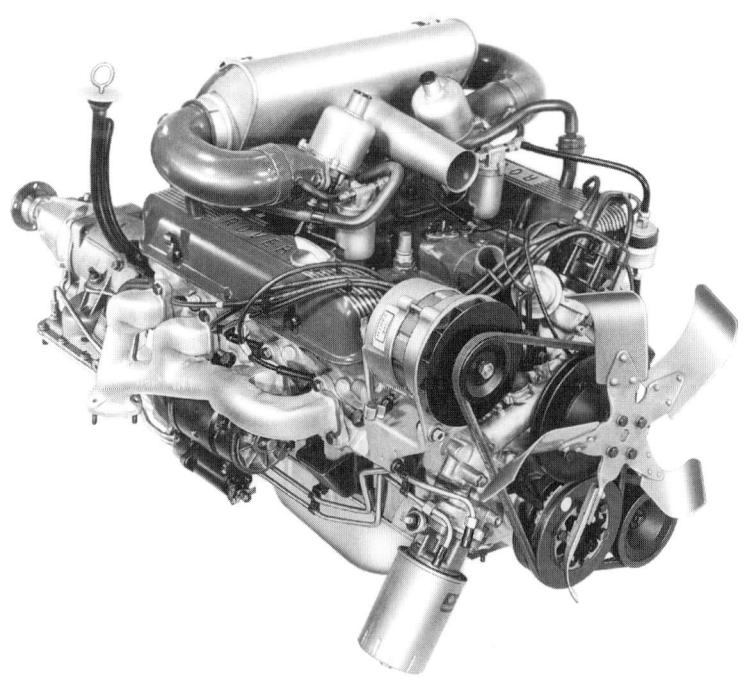

The Anglicized version of the Buick V8 engine first appeared in the Rover P5B. Its power output was eventually to surpass that found in the Vitesse.

and had a great deal of potential. Indeed, Jack Brabham's championship-winning Formula One cars in 1966 were powered by Repco engines based on the Buick V8 block.

The Rover V8 engine first appeared in the P5 model in 1967, and in the P6 model in 1968. To fit the new unit to the P5, it was necessary to add extra members inside the front subframe which carried the engine, transmission and front suspension. The P5B model (the B denoted Buick) was available in either four-door saloon or coupe form, initially as an automatic, as Rover did not have a suitable manual gearbox. The ageing P5 model was transformed with the new lighter engine – both performance (it was much faster than the three-litre) and roadholding were improved. For a while, P5 sales figures increased dramatically – after the 1967 Earl's Court Motor Show, twice as many cars a week were sold as Rover had anticipated.

The P6 model needed few changes to take the new V8 engine. The front crossmember had to be repositioned and the battery was moved to the car's boot. Some detail alterations had to be carried out. Early P6Bs were available as automatics only (the manual 3500S model finally appeared in 1971).

When the V8-engined version of the Rover SD1 first appeared in 1976, there were a number of engine changes compared with its predecessors. The power output was increased from 143 bhp at 5,000rpm to 155 bhp at 5,250rpm. The torque was slightly lower, at 198 pounds per foot at 2,500rpm instead of 202 pounds per foot at 2,700rpm, but engine flexibility was improved overall. The SD1 engine kept the 88.9mm bore and 71.1mm stroke, and capacity of 3,528cc. The camshaft remained the same, but changes to valves improved performance – the valving of the hydraulic tappets was changed to enable the engine to operate at slightly higher revs, and the exhaust and inlet valves were increased in size – exhaust valves were

Access to the V8 engine is not as cramped as this under-bonnet view of an early SD1 would suggest.

34mm in diameter instead of 33mm and inlet valves were 40mm diameter instead of 38mm. The valves also now had single valve springs. Airflow was improved behind the larger cylinder head valves. The compression ratio was lowered to 9.35:1.

The main difference in the SD1 V8 engine was the fact that the engine now breathed more freely at higher revs, owing to better designed, cast iron exhaust manifolds which were improved by having twin outlets. There was also a better water pump, and a different oil pump which gave a higher output at lower revs. A new electronic contactless ignition system by Lucas was another feature of the new SD1 engine.

The Rover V8 engine has been in production now for many years, testimony to William Martin-Hurst's good judgement. Its versatility and good torque meant that it could be used in off-road vehicles like Land Rovers and Range Rovers, where a wide torque band was more important than sheer power. Its performance has been developed and improved, not only by Rover themselves, but by numerous sports car manufacturers such as TVR. In the words of Spen King, 'the Rover V8 engine has been exceptionally successful'.

Birth of SD1 – key dates

Mar 69	Initial briefing for P10 (P6 replacement).
Mar 70	Initial programme for P10 in four/five-seat. Four-door saloon and four-seat, two-door sports saloon variants.
Jun 70	P10 package drawings passed to styling.
Feb 71	Project gathers pace – David Bache produces six scale clay models (five hatchbacks and a notchback) for consideration by the BL board. P10 project renamed RT1 – now also covered replacement of Triumph 2000/2500 (Innsbruck) models.
Apr 71	RT1 renamed SD1. (By this time the proposed four-door saloon and two-door coupe had been dropped in favour of a five-door fastback design with live rear axle).
May 71	First SD1 Features List prepared.
Jul 71	Interim (A) style approved by BL Board for production of A-batch prototypes.
Dec 71	Clay model of final (B) exterior style approved by BL Board for production.
Feb 72	Fibreglass 'see-through' model of interior and exterior style approved by BL Board.
Mar 72	All Rover A-batch drawings complete – work starts on production bodyshell drawings.
Jul 72	Later introduction of six-cylinder variants agreed.
Aug 72	Final B-style body skin information passed to body division.
Oct 72	Drawing programme for interior trim agreed, containing no slack and requiring early B-batch prototypes to be trimmed from schemes rather than from detail drawings.
Nov 72	B-batch fibreglass buck in use for cold room tests of SD1 heater.
June 73	Body tooling starts (aimed at pre-production in January 75 and pilot build in March 75).
Nov 73	Late drawing releases remain a problem – availability of optional extras and accessories under review in an effort to improve timing for the basic car.
Dec 73/Jan 74	Delays in some component supplies, resulting from the three-day working week, affecting the build of later B-batch prototypes.
Feb 74	Normal working hours resumed after three weeks of the three-day week emergency hours.
Feb 74	Pre-production postponed to mid-November and engineering sign-off to end-September.
Mar 74	Urgent programme in hand to evaluate a new central locking system which offered substantial cost savings.
Apr 75	First pre-production car built for further assembly methods development.
Sep 75	First pilot build car on production line.
Jun 76	Launch of 3500 model.
Oct 77	Launch of 2300 and 2600 models.

(Source: Mike Lewis, *British Leyland Mirror*)

3 The Birth – From P10 to SD1

The birth of the Rover SD1 project effectively took place at the time of death of the P8 project. The P8 was a Rover prototype which would have been a very large Jaguar-sized saloon with either an uprated version of the 3.5-litre V8 engine or a long-stroke 4.4-litre engine (this was eventually used in Leyland's Australian-made car, the P76). It would have been a very complicated car, with an advanced suspension system. The rear suspension was a development of the De Dion system used on the P6. Compliance for the front suspension was achieved by mounting it on a tubular sub-frame which could move all around on rods running through the tubes and bolted at the front to a body cross member and at the back to the member carrying the wishbone link pivots.

The P8 was fast, luxurious and intended to be 'a Mercedes eater'. As it turned out, it was too much like a Jaguar. P8 had been conceived when Jaguar and Rover were rivals. When both companies became a part of the Leyland empire, internal politics led to the project being cancelled, in March 1971. Although there was no carry-over of features from P8 to P10 (the original code-name for what eventually became the SD1), the warning signs for the future SD1 were clear to see. The Jaguar people in the Leyland Group had a great deal of influence. The new Rover would *not* compete directly with Jaguar – it would have to be aimed at the sector slightly below those people who traditionally bought Jaguars, slightly down-market. In this respect, then, there was a change of policy. Rovers before SD1, especially P4, P5 and some earlier models, had always been considered to be on a par with Jaguar cars. The Rover SD1, when it

The massive P8 prototype 'Mercedes eater' was killed off at a very late stage. The SD1's skeletal badge design first appeared on this car.

SD1 Project Personnel

Spen King (R) Director of Engineering and Product Planning

SD1 Engineering (up to Autumn 1974) – reporting to Mike Lewis (R), Chief Engineer with overall responsibility for SD1 project:
   Gordon Bashford (R) – SD1 Design
   Joe Brown (R) – Chassis
   Arthur Massey (T) – Body, Hardware and Trim
   Ray Fulbrook (R) – Electrical
   Peter Scholes (R) – Electrical
   Rex Marvin (R) – SD1 Development and Competitor Vehicle Evaluation
   Philip Mander (R)
   Eric Wright (R) – Engine and Transmission Forward Development (including V8 uprating)
   Richard Fishwick (R) – Research Section

SD1 Engineering (Autumn 1974 onwards) – reporting to Dick Oxley (R):
   Ken Stansbury (R)

Rover Styling – reporting to David Bache (R):
   Geoff Purkis (R)

Rover Purchasing – reporting to George Brown (R):
   Rover-Triumph Purchasing – Mike Fernyhough (RT)

Rover Quality – reporting to Ernie Bacon (R):
   Trevor Allen (R)

Pressed Steel Fisher Engineering – reporting to Charles Linder (PSF):
   Bill Emerson
   David Brown

Triumph Engineering
   Jim Parkinson (T) – Six-cylinder engine
   Mike Loasby (T)
   David Eley (T) – 77mm gearbox and rear axle

Rover-Triumph Production
   Alex Sanders (T)

Group Product Planning – reporting to Mike Carver (O):
   Alan Edis (O)

Key:
(R) = Rover origins
(RT) = Rover-Triumph origins
(T) = Triumph origins
(PSF) = Pressed Steel Fisher
(O) = Other

(Source: Mike Lewis)

---

came out, would be marketed against Fords, Renaults, Volvos and other motoring names which would have been considered less prestigious by the old Rover company.

Another factor which had a major impact on the SD1 project was the role of John Barber in the company. When John Barber joined the company from Ford he brought in a lot of financial people from Ford with him. He became Leyland's finance director in 1971. A costing department along the lines of that at Ford was set up, and the people in this department had a great deal of power. A number of people who were involved with the SD1 project, can remember that their hands were tied much of the time that they designed and developed the car. Every nut and bolt had to be accounted for, and the costs department was continually trying to reduce costs. This fact, of course, can be seen in the way the SD1 actually turned out – built down to a price. Rover engineers who had worked for the old Rover company had had far more freedom to carry out their designs.

The Birth – From P10 to SD1

The new Rover project, first called P10 in 1969, could not get under way properly until the P8 project had been killed off. After completing his project work on P8, Gordon Bashford's design team turned their attention to a smaller, new model, called P10. At first, this was purely a Rover project to replace the P6. There was a competition between Rover and Triumph engineers to produce ideas for the replacement of both the existing large Triumph and Rover models. Triumph came up with the Puma project. Both proposals (Puma and P10) were presented to the Board, and Rover won. After the cancellation of the P8 project, the P10 project was renamed RT1 (meaning Rover-Triumph project Number One) in February 1971 and became a joint Rover-Triumph project, to replace both the P6 and Triumph 2000/2500 models. By April 1971 the project was renamed SD1, meaning Specialist Division Number One.

Both Donald Stokes and John Barber were enthusiastic about the new Rover project. An early priority was to define exactly what SD1 would be like. In the words of Graham Robson (from *The Rover Story*):

> SD1, quite literally, had to be all things to all men ... At Rover it would replace .... all existing models. The big, old, but graceful 3.5 litre saloon would be allowed to die off ... and the P6 family would be replaced by a family of SD1 cars ... Eventually it would also have to replace the Triumph 2000/2500 cars, though these would carry on in production for a time at Canley after the first SD1s were built and sold.

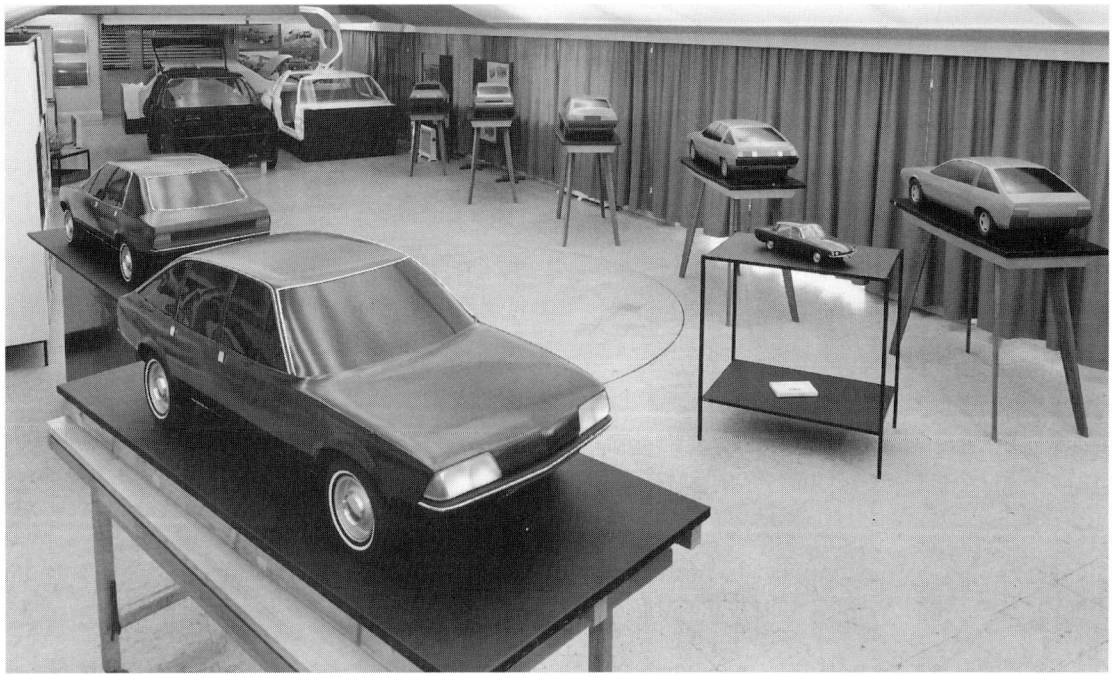

In February 1971, a quarter-scale model of the P10 prototype (in the foreground) was seen at a directors' preview, and compared with its Puma rival. There was also a full-size buck interior to show luggage capacity. At the rear of the picture is a mock-up of a proposed gullwing door mechanism.

The first styling drawings for the new car were extremely angular and not very inspiring, but the interior shows a strong Ferrari influence in the design of the seats.

The enthusiasm to make the new car a five-door, hatchback design came from stylist David Bache, and he was backed up by Spen King. A P6 hatchback prototype had been made previously – however, a new large hatchback design was a complete departure for the company. We are all familiar with large hatchbacks now, but at the time of the car's conception, when the largest hatchbacks were the Austin Maxi and Renault 16 (both of which were 1.5 litre cars), it was quite radical to design a large hatchback with a 3.5-litre engine. Spen King believed that the shape of the proposed new car lent

The Birth – From P10 to SD1

By the middle of 1971, David Bache had produced the first full-size mock-up of the interim (A) styling buck.

itself to the five-door, hatchback design. At one stage of the design, a quarter-scale model of a notchback was produced, but it never went beyond that stage.

Spen King remembers David Bache as being 'very much a law unto himself – you could not make him do anything he did not want to do'. The style of SD1 was pure David Bache. However, the shape of the car was no accident. His first attempts at styling were somewhat uninspiring – some of his early attempts were angular, and resembled the Chrysler Alpine. This Alpine-lookalike was an interim style to allow a few early prototypes to be built to gain experience on the body structure. Bache himself referred to this particular stage one styling of the new car as 'a railway carriage'. At this stage, he was very keen on the idea of gullwing doors. However, this idea never progressed any further than the mock-up stage, and was killed off. Gullwing doors were impractical, too complicated mechanically and would have been far too expensive to produce.

By July 1971, David Bache had completed his first full-size clay model – the A (interim) style. The Board thought this model too angular, and he was told to rework his ideas. David Bache became fed up with having his ideas rejected by the Board. At this point, he decided to do something radically different. He went off and decided to deliberately copy supercar designs, wanting the new Rover to look totally different from other large saloon cars. The car's styling was profoundly influenced by one of David Bache's favourite cars – the Ferrari Daytona. The front of the new Rover, with its front headlights and the styling crease along the side of the car came straight from the Ferrari. Indeed, at one stage of the design, one of Bache's styling attempts even had air outlets on its bonnet, like that of the Daytona. In a 1982 interview by *Car* magazine, David Bache admitted to the Ferrari influence on the Rover's design, and added that his own 1965 Rover-BRM gas-turbine car was also a source (this particular design having been influenced by the Ferrari 250LM).

Geoff Purkis, who was one of David Bache's styling team, remembers that as well as the Daytona, another car which influenced the exterior styling of the SD1 was the Maserati Indy, a four-seater supercar. Its influence can be seen particularly in the rear windows and the side profile at the rear.

In December 1971 David Bache produced a clay model of this sleeker, smoother shape – the B style. This was ratified by the Board in February 1972, and was the final form the Rover SD1 was to take. From that moment onwards, there were no further changes. SD1, in the shape that we now know it, was born. Its styling, without a traditional radiator grille, was dramatically different from that of any previous Rover model.

David Bache specially arranged a line-up of cars at the Solihull factory in 1972, including Maseratis, a Jensen Interceptor coupé and a large Mercedes saloon as well as more ordinary saloon models. These were placed alongside a styling model of the new Rover. The Maseratis were included to see how the Rover would stand out against supercars – it did not look out of place amongst such highly regarded company. As David Bache said (in *The Car of the Year, Today* brochure):

> We quite deliberately aimed at an exotic but long-lived style. Early clay styling mock-ups were put alongside cars like Maseratis and Ferraris. Despite the fact that it is a fully practical saloon car and not a cramped grand tourer, it looked perfectly in keeping.

By the middle of 1973, the B style prototype was well into testing – this is a three-quarter view of a road test prototype in disguise as a pick-up.

The Birth – From P10 to SD1

Some people found David Bache hard to work with – he fought passionately for his ideas and sometimes committed the engineers to solving almost insuperable problems in pursuit of them. However, Mike Lewis usually found it possible to persuade him to try another approach to any styling features which caused serious problems for the engineering department. David Bache himself said (in *The Car of the Year, Today* brochure):

> In a project such as the new Rover 3500 you find a degree of conflict between the various groups who are involved, but when we produce another vehicle, and if we have no more conflicts than we have had in this, I'd be happy. Most of the disagreements were on points of fine detail and all were amicably resolved in the end.

Contrary to popular belief, the somewhat stark interior styling on the new Rover was (unusually) *not* entirely the result of cost-cutting measures. The trim and interior styling details were the personal choice of stylist David Bache and his team, and were based on ergonomics and function. The Board adopted his concept of a simple elegant and perhaps understated interior, and in the design team's opinion this was achieved, at least on the interior styling bucks. Its effectiveness relied on good colour-keying, detailing and fits – Mike Lewis believes that possibly some of this was lost in production. He also feels that the simple style Bache advocated for the SD1's interior 'might now be seen as one of his rare errors of judgement', and thinks that engineering the trim scheme stopped before the detail finish on which it depended was fully realized. This possibility was forecast as early as 1972 because of difficulty in providing enough skilled trim designers. Trim and interior design were running late and not given sufficient time to be properly developed.

Geoff Purkis was involved with styling the car's interior. He remembers that the main influence here was the Ferrari Boxer

This line-up shows a styling model of the Rover SD1 between two Maseratis, and with other cars including a Jensen and a Mercedes. The exercise was so secretive that lorries were parked behind the line-up to act as a screen to prevent any unauthorized viewing.

The Birth – From P10 to SD1

Whilst looking very familiar, this picture of an interior mock-up shows a number of differences from the production car – apparently narrower luggage space may be seen, together with a different treatment of the front seat coverings.

– the shape of the seats, the door panels, arm-rests, and so on. Unlike the Ferrari's front seats (which were a single moulding), the Rover seats were manufactured in two parts to enable the fitting of the reclining mechanisms. Other features, such as the glove box, the fascia and instrument panel, and the air vent in the middle of the dashboard, were carried on from the P6.

The biggest problem stylists had was with the rear headroom, owing to the sloping roof line, and due to the amount of foam in the rear seats – there was a compromise with rear leg-room.

Bache refused to have a wiper on the rear window – he believed aerodynamics made it unnecessary. People involved with the project remember that a larger rear window became available just before the first SD1 came out, but this was too late to put into production for 1976.

Even the early paint colours on the new car (considered somewhat garish and un-Roverlike by some people) were chosen by Bache and his team, as was the non-circular steering wheel. The non-circular steering wheel was designed to enable the driver to see the large, circular instruments, and also to improve thigh clearance. The only adverse comments Mike Lewis can remember about the shape of the car's steering wheel were from the police, as this type of wheel did not suit their style of driving. There were, however, some qualms in the company about using a non-circular steering wheel on the new Rover; the Austin Allegro had received a poor reception for its pronounced 'quartic' wheel.

The new skeletal design for a Viking ship bonnet badge made in stainless steel was designed by Ian Beech, and approved by David Bache. Ian Beech had a background in the silversmith industry – it was no surprise, then, that his design drew on the influences of Scandinavian jewellery.

Originally, it was envisaged that the instrument panel would be modular, and made up of individual parts. The design

The SD1 design team

Spen King

Charles Spencer ('Spen') King joined Rolls-Royce at Derby as an engineering apprentice during World War II. Here he became involved in the design and development of jet engines, joining Rover in 1945 and working on the conception of the Rover 'jet' cars – the gas-turbine cars – until 1961. He then moved over to Rover's production cars, and was closely involved in the design of the Rover 2000 (P6) range of cars. Spen King went on to mastermind the Range Rover project, and together with Gordon Bashford, produced the P6BS coupé. This car was a mid-engined sports car, and would have been produced by Rover if the newly-formed British Leyland had not abandoned the project. He was also involved with the development of the Triumph Dolomite Sprint and the Triumph Stag. Spen King was director of engineering and product planning at the time of the SD1 project. He was enthusiastic and had a 'hands-on' approach to his work, and enjoyed sorting out problems in the development workshops. When BL was broken up, Spen moved on to BL Technology, which was set up as an advanced engineering and manufacturing group. He left the company in 1985, to set up an engineering consultancy. He is regarded by many people as one of the great chief engineers of all time. Spen King lives in Warwickshire, and his hobbies include a passionate interest in sailing and also photography.

Mike Lewis

Mike Lewis joined the Rover company in 1954 as a graduate apprentice. He joined the engineering department the following year, and worked on the torque converter for a semi-automatic transmission for the P4 model. Mike Lewis joined the company's research section when it was formed two years later, and was promoted to chief research engineer in 1966. He was appointed chief engineer, new vehicle projects, in 1971 with overall responsibility for the SD1. In autumn 1974, before pre-production started, he was asked to hand over responsibility for engineering SD1 to the current cars department and to take up a more senior position in the emerging combined Rover-Triumph engineering department. It is clear that the SD1 project formed a major part of his life. Later on, he was the brains behind the formation of BL Technology. Nowadays, Mike Lewis lives in Cornwall, and his hobbies include an interest in 'God's Wonderful Railway' – the Great Western.

David Bache

David Bache trained at Birmingham University and College of Art, starting his motor industry career as a student engineering apprentice at the Austin Motor Company. One of his earliest projects here was the design of the instrument panel for the Austin A30. There were so few competent designers in the Midlands at this time that he was headhunted by Rover, and he joined the company in 1954. Sent to the Paris Motor Show, he was profoundly influenced by the new Facel Vega. This model's sleek, handsome lines were reflected in a series of Rover cars styled by Bache. The first Rover car to be styled by David Bache was the three-litre P5, followed by the Mark II Land Rover. However, his first important design was the Rover 2000, which carried the unique sculptural stamp which characterized all of his work. His body design for the gas-turbine Rover showed the successful combination of high style with competition aerodynamics. Later designs included the P6BS project and the Range Rover. At the time of the SD1 project, David Bache was Rover's chief engineer of styling, and was responsible for the car's unique shape – making sketches, silhouettes and then having clay models produced. The whole SD1 project was a labour of love for him. After SD1, he became Leyland Cars' director of styling, and later went on the shape both the Metro and the Maestro models, retiring from the company in 1981. He was once described as the most influential designer in the British Motor Industry. David Bache died in 1994. Appropriately, there was a line-up of the Rover cars he had styled outside the church at his funeral service.

Gordon Bashford

Gordon Bashford had an incredibly long career with Rover. He contributed a great deal to the

The SD1 design team

marque, being involved in the design of all of Rover's models from the 1930s to the SD1 of the 1970s. He joined the company in 1930, straight after leaving school. A friend introduced him to Roland Seal, who was at that time the head of the drawing office at Rover. Seal gave him a junior job in the department paying the (then) princely sum of ten shillings a week. The very first car Gordon Bashford was concerned with was the Scarab – an advanced, rear-engined car. However, this project was eventually abandoned and the Scarab was never put in to production. He became involved in the design of the P1 models – these cars acquired a good name for the company during the war, when they were able to keep going with the minimum of maintenance. After his roles in designing both the P3 and P4 models, Gordon Bashford was asked to tackle the layout of the Land Rover. He had no previous experience of designing four-wheel-drive vehicles, but got this task just right – the basic Land Rover layout remains unchanged to this day. He was also involved in the designs of the P5 and P6 models, and the P6BS and P8 prototypes, and worked very closely with Spen King on both the gas-turbine cars and the Range Rover. Gordon Bashford was responsible for the SD1's design and he believed that there was a good deal to be said for the model's live axle rear suspension system. After the SD1, he made a substantial input to the chassis engineering of the Maestro. His hobby was designing and building racing cars in his spare time. Gordon Bashford retired from work in 1981, and died in 1991. He is fondly remembered by all who knew him.

The Rover SD1 design team were named 'Midlanders of the Year'. Left to right: Mike Lewis, David Bache, Spen King and Gordon Bashford.

The Birth – From P10 to SD1

team could not carry out this idea in practice – the problem was making everything line up without creaking all the time when the car was in motion. However, the appearance of a modular assembly was retained on the actual production car as a styling feature. David Bache wanted the car to have an electric speedometer, which could be easily removed for servicing – however, this was rejected on cost grounds (it would have cost £2 extra per car). For impact safety, the fascia was formed as a crushable steel pressing with protective foam.

There were some concerns about the absence of wood and leather in the new car. The first item, the lack of wood, was an extension of the trend between the earlier P5 and P6 models. Leather seats eventually became an option later on in the production life of the car, when problems with maintaining their initial appearance had been solved by the designers. The velour upholstery for the car was a specially-developed, durable nylon material.

David Bache put a great deal of effort into designing the driver's seat. It tilted in three ways to accommodate short, tall or average height drivers, as well as giving a good range of seat back adjustments. Engineering did have some problems with the seat design. It had originally been intended that the front seats should have pressed steel frames. However, in July 1973, at a late stage of the project, they had to use tubular frames in the front seats, due to tooling costs going over-budget.

Bache's styling did cause some problems for the engineers. His original shape had started out 1.5in (380mm) shorter than the car eventually ended up in production. There was interaction between these two departments to sort out any problems. Some changes had to be made to the exhaust system David Bache had originally envisaged, because exhaust fumes got into the car. The sill height was determined by the styling, as was the amount of headroom.

With regard to the engineering of the new car, keeping down costs was of overriding importance. Design requirements were for a refined, safe car which handled well. It would be versatile, and offer value for money, with a number of items, traditionally regarded as optional extras, which would be fitted as standard. From the start, the whole of Europe was seen as the home market, and restful, economical high-speed cruising on motorways was seen as an important requirement. (For this purpose, extensive wind tunnel tests were carried out, to improve the aerodynamic performance of the new car and to improve fuel economy.) It would need to be cheap and easy to manufacture and assemble. The new Rover would have a monocoque body, which would be light, with welded panels – this would be more corrosion-proof than the P6 had been, with its bolted on panels.

The Rover SD1 was possibly the last Rover to be designed in the traditional way, on the drawing board. There were no computer aided design facilities at British Leyland at this time. However, Pressed Steel Fisher, who carried out the bodyshell design from Rover's clay models, used computer aided design to design the body skins and structures, and for press tool design. After drawing the shell on computer, Pressed Steel transferred the design back to Rover for trim and hardware design. Computers were used for figure analysis and so on (for example, when David Eley's team were designing the transmission components, computer programs were used to help select suitable proportions for the gear teeth).

Rex Marvin, who was responsible for developing the new Rover, remembers that at the very beginning of the project, the design team were very keen for the SD1 to be a front-wheel drive car. There was concern over rapid wear of the front tyres, owing to the large braking forces on the front wheels, and it was thought that this problem could be alleviated by the use of ABS. No prototypes or components were manufactured, although a BMC 1800 had its braking system modified in such a way that the braking effort could be varied in proportion between the front and rear wheels, in order to gain an impression of the feasibility of the idea. The concept of front-wheel drive was then rejected – the design team believed that it did not offer much in the way of packaging advantages on a medium to large car over a conventional rear-wheel drive layout which was known to be able to give good handling characteristics – something that is not always present with front-wheel drive. Front-wheel drive would also have possibly been a more costly option for the company to put into production at this time.

Therefore, early on in the project, it was decided to design a mechanically simple car with a live rear axle, with none of the complications of its P6 predecessor. A Peugeot 504 estate car was used to evaluate what a live axle was like on a car of this size; it was found to be acceptable. A lot of work was carried out evaluating the rear suspension, using a P6 3500 with a Vauxhall Ventura live axle instead of the P6's usual De Dion arrangement, and Boge suspension like that used on the Range Rover. This vehicle was taken all over Europe for testing purposes. It was only after riding in this vehicle that Spen King (creator of the much-acclaimed rear suspension for the P6) accepted that the live axle gave an acceptable ride.

With regard to the suspension design, Mike Lewis would have liked the car to have had a separate tie-bar to provide longitudinal location of the front wheel assemblies, the anti-roll bar providing roll stiffness only and not any contribution to the geometry of the suspension system, allowing the suspension bushes to be designed solely to provide lateral stiffness and fore-and-aft compliance. Preliminary costings suggested that there was little to choose between the two schemes. However, the integral anti-roll bar performed well on both functional and durability tests once problems with its sideways location had been solved and after it had been thickened to cope with suspension loads; the simpler scheme was therefore adopted. Lewis was not aware of any problems of premature wear in normal use with the suspension bushes specified for production, but feels that the alternative scheme might have been more tolerant of bush material changes. (It could possibly have alleviated the problem of premature wear of suspension bushes and the overstressing of the anti-roll bar mountings.)

When developing the SD1, a number of different vehicles were used to test the new car's suspension, steering, back axle, front structure, heating, V8 engine and so on. The six-cylinder engine and gearbox was tried out on a Triumph 2500 to help evaluate and develop the gearchange quality, and to see how it performed on the road.

P8 prototypes from that abortive project also came in useful for testing purposes. The P8 had similar track and wheelbase dimensions to that of the proposed SD1, and in order to compensate for the P8's heavier weight, some of its outer panels were replaced by panels moulded in fibreglass. These P8 cars were mainly used for testing the cooling and suspension systems, and so on. Even a P5B was used for gearbox and axle development. Various MacPherson struts were tried out on different cars.

During development, designers tried out a rear axle located with a Panhard rod as an alternative to the Watts linkage (to cut costs if possible) but this was found to be functionally inferior and thus rejected. Therefore, the actual chassis engineering of the SD1 was fairly conventional, apart from its self-levelling suspension system. Later on in the development of the new car, the team found that after certain modifications to the Nivomat rear suspension units to overcome ride interference problems, the ride of the new SD1 compared favourably with Mercedes 230 models.

Mike Lewis particularly wanted the new Rover, though big, to be a car which could be parked easily – a turning circle no larger than that of the P6 was therefore specified, and was maintained in spite of a further wheelbase increase during development to allow more space for the engine sump. The SD1 had responsive, quick and predictable steering, and a good turning circle of 10.0m compared with 10.8m for its predecessor, in spite of the longer wheelbase. This made the car easier to manoeuvre in narrow streets and tight parking spaces, in spite of its size.

Prototypes were extensively tested, in Britain, Europe and North America. The cars were taken to Northern Norway for cold climate testing, and to Nevada for high-speed endurance and hot climate testing. Rex Marvin and Phil Mander were both involved with testing the prototypes overseas. The cars were taken to Nevada because of its hot climate, high altitude and the fact that it was one of the only places you could do 130mph in a straight line. Tyres tests were carried out in North America, including on Denovos. On one occasion, while testing the car, one of the wheels with its half-shaft was jettisoned at speed, and a fire broke out – this was not a design failure, but an assembly failure due to lack of lubrication. Another time, they took a prototype to the far north of Norway, having driven it there from Hammerfest. The car and its heater performed satisfactorily. However, after they had arrived at their hotel late at night, they discovered a convention of Volvo dealers staying there; they had to sneak away before dawn.

The front structure of the SD1 was consciously modelled on the Triumph 2000. Consulting with the British Insurance Association's Research Centre at Thatcham, the designers decided that the new car would have welded panels, which would offer the advantage of being lower in weight than a car with bolted-on panels, and were satisfied with the repair implications for owners. Mike Lewis's team had considered using foam-filled sills, but consultants at Thatcham advised them that this would not be a good idea – they had found that noxious fumes could be given off if the sills were repaired by welding. Therefore, the idea was abandoned. The SD1 was designed to avoid moisture traps leading to corrosion, by means of forced ventilation fed from the heater plenum chamber for the sills, the outer skins of which were made from zinc-plated steel.

The design team made a conscious effort to overcome the corrosion problems found in the P6. To this end, they introduced the novel concept of positive air-flow through the length of the body sills – a feature which time has proved worked.

The Birth – From P10 to SD1

Great effort was put into making the Rover SD1 a safe car. This picture shows a Rover 3500 after a MIRA one-third overlap impact test into a rigid barrier at 30mph.

Great emphasis was placed on designing a safe car (the P6 predecessor had been noted for its safety features). A number of prototypes were devoted to various forms of crash-testing, covering both current and emerging legislative requirements around the world (for example, US 40mph front barrier impact, rear barrier impact and body roll-over tests), as well as the company's own in-house safety requirements. These crash tests included the standard 30mph full barrier front impact, side and rear impact and roll-over protection. Early 30mph barrier impact tests on the cars revealed problems with the MacPherson strut pulling away from the car body, and some modifications had to be carried out to the extreme front structure to improve energy absorption.

One in-house requirement which was not legislated for was that all doors of the car should open without tools after the front impact test – designers ensured that this was possible by having heavy tubes inserted in the doors to distribute safely the resulting compressive loads of the crash. An offset frontal impact test (also unlegislated) was done to allay concerns over the effectiveness of the collapsible steering column. Also, tests were carried out on the seat belt anchorages, to ensure that they did not pull out of the body frame of the car. Front side window demisting and front door edge safety lights were added to improve safety.

A Pirelli long-term tyre testing team accidentally drove an early SD1 through a massive brick wall and emerged from this experience unscathed. Unsurprisingly, they praised the integrity of the Rover SD1. The inevitable road accidents with early vehicles gave the designers some confidence about the car's safety features. As Spen King has remarked, the SD1 cars are pretty good in an accident.

Numerous modifications were made to reduce production costs. For example, when it was found that the proposed open cable and pulley handbrake linkage would be more expensive than twin enclosed cables, the latter were specified for production. A new bumper with rolled centre section and urethane end pieces was designed when the purchase department found they could not source the conventional one-piece design competitively. The type of central locking system chosen was arrived at partly because it offered a substantial cost saving over the previously envisaged vacuum system. Prototype headlamps were designed by both Lucas and Cibie, but the Lucas ones (which had not worked so well in testing) were eventually chosen on cost grounds. The drum brakes on the rear axle were adequate, but were chosen mainly because they were cheaper than disc brakes.

David Eley was responsible for designing and developing the new five-speed gearbox for the car, and also its rear axle. He worked closely with quality assurance and production people. The 77mm gearbox was named after the spacing between the mainshaft and layshaft centres. It was a brand new design, and was unusual for two reasons – it had tapered roller bearings, as opposed to conventional ball bearings (which gave it a much higher load capacity), and a pumped oil feed to the layshaft assembly. The back axle was of conventional design with a long extension shaft for the pinion housed within a torque tube (to provide part of the suspension location). Its chief novelty was the fitting of a single special tapered roller bearing at the outer end of each half-shaft (this being an innovation by Timken, the American firm, known as a unit bearing, which, unlike a conventional tapered roller bearing, could provide a limited amount of axial location in the opposite direction to what is normal, thereby allowing the use of one bearing only in each hub).

The new gearbox had to be designed to cover a range of cars, including the TR7 and more powerful cars such as Jaguars. This factor influenced a number of features of the gearbox design. For example, a generously proportioned clutch assembly was chosen in order to cope with the high-powered engines envisaged. One Yorkshire police force got 30,000 miles life per clutch – this was considered remarkable, and David Eley believes this was because 9.5in (240mm) clutch plates were used. (This particular police force used their SD1 cars for 105,000 miles before replacing them.)

The reason tapered roller bearings were chosen was that their load-carrying capacity was that much greater than ball bearings and bushes that would otherwise have been used. This was brought about by the requirement to make the gearbox as compact as possible in order that it could fit into the range of vehicles, some of which would be smaller than the SD1. The use of tapered roller bearings necessitated the main gearbox castings having to be made of cast iron (the designers had originally wanted to use aluminium). Cast iron has similar thermal expansion characteristics to the internal working parts, thereby allowing the gearbox to operate freely over a wide range of ambient temperatures. In this instance, the choice of material was not arrived at on cost grounds.

Because of the requirement to use the SD1 gearbox with high-powered engines, it was felt prudent to provide forced feed lubrication of the gears with an oil pump driven from the layshaft. As it was envisaged that the transmission would be running in fifth gear much of the time under conditions of high power (such as in motorway use), the indirect fifth gear was provided with needle roller bearings.

Gearbox problems were solved early on. Although they worked adequately on the test rigs, a few gearboxes failed in service, which resulted in a modification which was quickly put into production. A stress raiser near the end of the layshaft where the fifth-speed gear was mounted was dispensed with. This cured a weakness, removing the wasted area from the shoulder of the shaft that the fifth-speed gear located against. This had the advantage, too, that it made the gearbox easier to make. Another problem was the not uncommon one of the workforce not following the drawings closely enough, leading to tolerance problems. Also it was discovered that oil viscosity had a marked effect on the gear change quality in cold conditions. When it was cold, thick oil caused problems with drag and longevity – this problem was cured by using lighter oil and by ensuring greater accuracy of shimming to control excessive pre-loading. This brought about the unusual choice of automatic transmission fluid for use in the manual gearbox.

Extensive endurance testing was carried out of both the gearbox and the back axle, with test rigs running for twenty-four hours a day. The BL plant at Pengum in Wales was responsible for the assembly and testing of transmission components; the back axle was made and tested at the BL plant in Coventry. Progress on the new gearbox proceeded far quicker than that made on the new six-cylinder engine.

The design team never envisaged that a limited slip differential would be needed on the SD1, therefore this idea was never tested. The designers felt there was no requirement for this item, and that the limited slip differential is never totally satisfactory on normal road cars.

The biggest problem encountered during the whole SD1 development programme concerned the propshaft and petrol tank. It had been decided for safety reasons to mount the petrol tank ahead of the rear axle and to one side of the propshaft. To give a tank of reasonable capacity, it had been found necessary to offset the torque tube of the axle. The resulting angled line of the propshaft required the constant velocity joints to run continuously at a considerable angle. The makers, GKN, had rig-tested the propshaft assembly both for maximum offset and maximum power and had found it satisfactory, but unfortunately had not tested it with both these maximum parameters simultaneously. This was dramatically revealed in subsequent road-testing, with the rubber boots bursting open explosively. (This was the result of the heat being generated by the working of the joints, with vaporization of the lubricant, this generating enough pres-

An example of the 77mm gearbox sectioned for exhibition purposes, showing the unusual construction of its casing with the cast iron central section, necessitated by the use of taper roller bearings.

sure to rupture the sealing boot, with consequent smoke and mess.) The position of the propshaft and petrol tank meant that the handbrake had to be sited slightly away from the driver on home-market cars.

The depth of the petrol tank was severely restricted by the location of the floor under the rear seat and the ground clearance. Eventually, this problem was resolved by settling for a fuel tank of reduced size; this also allowed the offset of the driveline to be less extreme. The resulting 14.5 gallon (66 litre) fuel tank was small for a car of this size.

These were the most serious problems engineers had to solve on the project, fortunately occurring fairly early on and therefore rectifiable. Conventional Hooke-type joints were tried out on the propshaft when testing, but these were eventually rejected because they were not refined enough, the car suffering from an unacceptable amount of vibration.

The automatic Borg-Warner unit had been subjected to continuous development and design improvements. The change from the Borg-Warner automatic transmission to the General Motors one later on in the production life of the SD1 did not come about through any technical shortcomings of the Borg-Warner transmission, but was carried out in order to ensure continuity of supply.

The six-cylinder engine underwent a painstaking design and development programme, which started in the very early 1970s. It finished up as 'a clean sheet of paper' design, with no parts in common with its Triumph predecessor. Like other areas of the SD1's design, the team who designed the new engine were very constrained on costs.

Mike Loasby, who was involved with designing the new engine, joined the Company in 1969. Immediately previously, he had been a development engineer at Aston Martin, and before this had worked for the Coventry Climax company. At the time of SD1, he was an engine designer for Triumph and Rover. After the SD1 project, he returned to Aston Martin as chief engineer.

The main problem with engine design is getting the breathing right. To this end it was decided at the beginning of the project that the engine would embody overhead camshaft valve gear instead of pushrod-operated overhead valves, as on its predecessor, as this layout ensured that the valve timing accuracy was much more consistent and there were no pushrods in the way of the valve ports, which could be of generous size. The engine design had pent-roof combustion chambers with inclined valves. In order to maximize the sizes of inlet and exhaust ports, the valves were not directly opposite each other, but staggered along the length of the engine. Whilst only having two valves per cylinder, the valve gear followed the general lines of that embodied in the Triumph Dolomite Sprint engine, which had combustion chambers with four valves each.

One of the problems in engine development is of controlling oil flow in the cylinder head area. If there is too much oil it will tend to lead to a high oil consumption, with the oil passing down valve guides. Therefore a restrictor was put in the cylinder block joint face in order to allow a drilled oil passage of reasonable proportions to be made to the lower part of the engine. An additional idea was to have intermittent feed to the valve rockers, rather than continuous feed, this idea having been used on previous car models.

The six-cylinder engine had a number of innovative production features, including the use of powder metal technology in the manufacture of the combined crankshaft timing pulley and torsional vibration damper (which would have been impossible to produce by orthodox machining methods). The water pump impeller was also a sintered item which allowed the combination of efficient functional form and the complete elimination of machining processes. The oil pump and water pump housings were die-castings with little or no machining. The fixing of the exhaust manifold to the cylinder head without the use of a gasket was also unusual and would only be possible if the castings were designed to maintain rigidity with machined surfaces of the joint being finished to a high standard.

The cylinder block was rigidly designed. This had the innovation of much webbing and corrugation to help with the general refinement of the engine. The whole block was conceived in such a way as to simplify the casting process, to minimize the number of cores in the sand mould into which the liquid iron is poured, thereby improving the quality of the casting, resulting in higher accuracy – this improved quality and lowered production costs.

The cylinder block was originally designed with a water jacket of minimal depth, as it was felt that this would be an aid to manufacturing the casting. However, this arrangement was found functionally unsatisfactory, and in practice the manufacturing problem with the casting at the foundry did not exist, so the final version of the cylinder block had a much deeper water jacket, with no cooling problems.

So far as possible, oil ways were cast, as opposed to drilled, which had been orthodox practice previously (a possible source of production difficulties, and a likely trap for swarf and dirt).

In designing the cylinder head the designers were bearing in mind North American anti-pollution regulations and exhaust gas recirculation for future sales in the USA (this is ironic, in view of the fact that the six-cylinder SD1 was never sold there).

Control tightening of bolts was employed for the cylinder head joint on the production line, which eliminated inaccuracy in assembling the cylinder head gasket. Because of the limited space around the cylinder head bolts, the design team used a multi-splined bolt rather than an orthodox hexagon head. In order to reduce costs, it was at first thought that the expensive helical gears used to drive the distributor could be dispensed with by mounting the distributor horizontally at the front end of the cylinder head and driving it directly from the camshaft. At this time, no engine was in production with such an arrangement, and concern was felt for the possible leakage of engine oil into the distributor. It was also felt that such an item might be in the way of the downward-sloping bonnet line.

The traditional material of a steel forging was used for the crankshaft, rather than a nodular iron casting (because steel is a much stronger material, and is capable of withstanding the high power outputs of which this engine was potentially capable).

In a typical car engine, the connecting rods have to be individually weighed and selected in matched sets to ensure the correct engine balance with the resulting refinement. In the new six-cylinder engine, this process was eliminated by ensuring that about 80 per cent of the surface area of the connecting rods was machined, (mainly around the big-end eye).

To ensure that the connecting rods were as strong as possible within the constraints of their overall size and weight, it was necessary to have the big-ends split at right-angles to the rod axis. This fact determined the minimum diameter of cylinder through which the connecting rod would pass during assembly (the engine had a bore of 81mm).

At first, Beans Industries were opposed to Mike Loasby's crankcase design, and requested simplification. The design broke a lot of new ground with the foundry. The water jacket core was made in one piece. The original intention was to make the two blocks back to

The Birth – From P10 to SD1

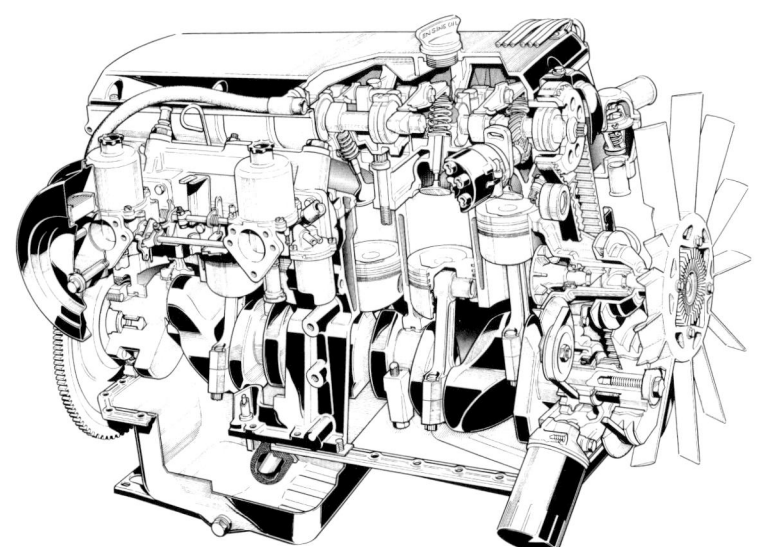

Cutaway drawing of the six-cylinder engine as fitted to the 2300 and 2600 cars, showing its generally massive construction, despite which it has established an unenviable reputation.

back, (looking somewhat like a horizontally-opposed twelve-cylinder engine when in the mould), although this idea was not put into production. In addition, the foundry was made responsible for some initial machining to ensure the block castings could be accurately finished to size on transfer line machines. (Previously, foundries had only provided rough castings for motor manufacturers to machine, with the possibility of having to reject malformed castings.)

It was originally intended to replace the old Triumph engine with one new engine of 2,300cc. However, the designers later decided to add a 2,600cc engine. The engine was basically the same for both models, but the 2600 had a longer stroke (84mm instead of 76mm), and apart from different crankshaft and pistons, there was little difference between the two engines – the 2600 engine was fitted with high-duty Mahle strutted pistons with solid skirts, and the combustion chamber partly in the piston crown, whereas the 2300 had Hepworth and Grandage W slot-type pistons.

The design team gave detailed consideration to an aluminium engine block, but no prototype was ever constructed. (Aluminium was possibly rejected both on cost grounds, and because it may not have been as refined as the iron block engine.) Early on, it was decided to put the six-cylinder unit into the Rover SD1 only, and not to use it in any Triumph models. A couple of prototypes were tried out in Triumph 2.5 PIs, for experimental purposes only.

All production engines were run for a few minutes on the test rigs. Initially, twenty-five prototype engines had been made – with possibly a few hundred cylinder heads. Early six-cylinder prototypes had more power than the V8 engine when they were first run – the 2600 engine could easily have produced more than 150bhp, but power and torque were deliberately kept down so as not to outperform the V8 version of the car.

There were a number of problems to solve in the development of the six-cylinder engine, among them being difficulties with piston rings, which in part were overcome when it was decided to extend the water jacket downward, which had the side-effect of reinforcing the cylinder bores (there is always bore distortion and cylinder head distortion when developing new engines). The engine which was eventually used was the third version to be considered. When it was first road-tested, it ran smoothly but suffered from the low-speed torque, which led to the fitting of a different camshaft to improve performance.

In comparison with the design and development of the new six-cylinder engine, progress with the existing V8 engine was relatively straightforward. Maximum power endurance tests were completed satisfactorily. Relatively minor problems such as fatigue failure of the oil strainer assembly were soon put right. For the new Rover SD1, the V8 engine had undergone further development and improvement. Compared with previous versions of the engine, the SD1 engine had larger valves, revised valve timing (different camshaft) and better oil seals on the crankshaft.

There are always problems to sort out when a new car is put into production, especially if it is significantly different from its predecessor and if the manufacturing plant is also new and unfamiliar, as was the case with the SD1. In an effort to reduce the number of problems with the new car, production engineers worked alongside the designers. Also, efforts were made to reduce the number of component suppliers so that each could be fully involved in the project. It had been originally intended that the SD1 engineering team should stay with the car until it was established in production; this had not been done with some earlier Rover projects. Mike Lewis feels that the engineering hand-over in the autumn of 1974, caused by the imminent amalgamation of the Rover and Triumph engineering departments, was premature and contributed to some later difficulties in finalizing the design for manufacture. He mostly enjoyed the personal challenge of the SD1 project, and was disappointed not to have been able to carry the car through to production.

A number of people involved with the development of the SD1 believe that there was too much concentration on developing the car for the North American market, at the instigation of product planners. With the benefit of hindsight, British Leyland should have concentrated on getting the car completely right for the home market first.

A number of industrial problems helped to cause delay to the whole SD1 project, from the design stage onwards. A dispute in the drawing office at Pressed Steel Fisher delayed the A-batch prototypes. Later on in the project, the three-day working week delayed the B-batch cars. There were also some problems in keeping up overtime working because of the pressure on pay and conditions, and some difficulty in retaining staff because of statutory pay restraint – these problems were particularly acute in the trim design department.

The six-cylinder engine programme (the cars had originally been intended to be launched very soon after the 3500 version) was particularly badly delayed. There were several reasons for this. All of the prototype engine blocks cast by Beans Industries were supposed to be machined by by a firm which went bust in the middle of the job. There was also a problem with the introduction of the early production machinery while development was sorted out. Parts were delayed, and the timing of the whole project was thrown off course.

In spite of all these problems, body tooling for the new car started in 1973, and the first pilot built car was on the production line by 1975. The 3500 model was ready for launch in 1976, with the six-cylinder models launched a year later.

The SD1 prototypes

Date to sales	Vehicle Identification No.	Notes
10. 6. 75	RRWVF3AA000001	Engineering development
24. 6. 75	RRWVF3AA000002	Engineering development
15. 9. 75	RRWVF7AA000003	Engineering development
10. 6. 75	RRWVF3AA000004	Engineering development
24. 6. 75	RRWVF3AA000005	Quality control
24. 6. 75	RRWVF3AA000006	Engineering development
8. 7. 75	RRWVF3AA000007	Engineering development
10. 7. 75	RRWVF7AA000008	Engineering development
10. 7. 75	RRWMU3AA000009 (no engine)	Engineering development
18. 8. 75	RRWVF3AA000010	Engineering development
1. 12. 75	RRWVF7AA000011	Quality control
15. 9.75	RRWMU7AA000012 (no engine)	Engineering development
15. 9. 75	RRWVF3AA000013	Engineering development
15. 9. 75	RRWVF4AA000014	Methods build
25. 8. 75	RRWVF4AA000015	Engineering development
25. 8. 75	RRWMU3AA000016 (no engine)	Engineering development
25. 8. 75	RRWVF3AA000017	Engineering development
15. 9. 75	RRWMU7AA000018 (no engine)	Engineering development
15. 9. 75	RRWMU3AA000019 (no engine)	Engineering development
1. 12. 75	RRWVF8AA000020	Engineering development
1. 12. 75	RRWKA4AA000021 (no engine)	Engineering development
1. 12. 75	RRWVF8AA000022	Engineering development
1. 12. 75	RRWVF7AA000023	Engineering development (Richelieu/Coriander)
1. 12. 75	RRWVF7AA000024	Service training, Allesley
1. 12. 75	RRWVF8AA000025	Technical publications, Allesley
10. 9. 75	RRWVF3AA000026	Methods build
1. 12. 75	RRWVF3AA000027	Technical publications, Allesley
1. 12. 75	RRWVF3AA000028	Marketing, Allesley
1. 12. 75	RRWVF3AA000029	Marketing, Allesley
1. 12. 75	RRWVF7AA000030	Quality control
1. 12. 75	RRWVF3AA000031	Press department
4. 12. 75	RRWVF3AA000032	Press fleet
4. 12. 75	RRWVF3AA000033	Service department, Allesley
1. 12. 75	RRWVF3AA000034	Press fleet
1. 12. 75	RRWVF3AA000035	Press fleet
21. 10. 75	RRWVF3AA000036	Press department (film unit)
1. 12. 75	RRWVF8AA000037	Marketing services (photography)
16. 3. 75	RRWVF4AA000038	Engineering development (homologation)
1. 12. 75	RRWVF3AA000039	Press fleet
1. 12. 75	RRWVF3AA000040	Press fleet
21. 10. 75	RRWVF3AA000041	Press fleet
21. 10. 75	RRWVF3AA000042	Press fleet
1. 12. 75	RRWVF3AA000043	Press fleet
1. 12. 75	RRWVF3AA000044	Press fleet
1. 12. 75	RRWVF7AA000045	Press fleet
1. 12. 75	RRWVF3AA000046	Press fleet
1. 12. 75	RRWVF3AA000047	Press fleet
1. 12. 75	RRWVF3AA000048	Press fleet
1. 12. 75	RRWVF3AA000049	Press fleet
1. 12. 75	RRWVF7AA000050	Press fleet

(Source: British Motor Industry Heritage Trust)

4 The New Rover 3500

Old and new models compared		
	Old 3500 (P6B)	New 3500 (SD1)
Wheelbase (in)	103.5	110.5
Length (in)	179	185
Width (in)	66	69
Height (in)	56	53.5
Weight unladen (lb)	2,872	2,895
Front track (in)	53.5	59
Rear track (in)	52	59
Fuel tank capacity (gallons)	15	14.8
Tyres	185-14	185-14
Payload (lb)	917	1,245

(Source: *Autocar*)

The 3500 model of the Rover SD1 was launched on 30 June 1976, and was widely acclaimed by both the press and the general public.

It was a very different beast from its P6 predecessor. Performance was not vastly different between the new car and the high-performance version of the P6, the 3500S (0-60mph in around 9 seconds and a top speed of just over 120mph). However, in other respects the two cars were like chalk and cheese – the SD1 was a relatively simple design mechanically, when compared with its P6 predecessor. The new 3500 was 7in (178mm) longer in wheelbase and offered more legroom, but was only 6in (152mm) longer overall. It was 3in (76mm) wider but 2.5in (64mm) lower – the

The result of all the foregoing – the new Rover 3500 as announced in 1976.

Rover 3500 (1976–86) – specification

Engine
Cylinders	8, in 90 deg. vee
Main bearings	5
Cooling system	Water
Fan	Viscous
Bore, mm (in)	88.9 (3.50)
Stroke, mm (in)	71.1 (2.80)
Capacity, cc (cub in)	3,528 (215)
Valve gear	ohv hydraulic
Camshaft drive	Chain
Compression ratio	9.35 to 1
Octane rating	97 RM
Carburettors	2SU HIF6
Max power	155bhp (DIN) at 5,250rpm
Max torque	198lb ft at 2,500rpm

Transmission
Type: Five-speed manual, 9.5in SDP clutch; Borg-Warner Type 65 automatic extra

Gear	Ratio	mph/1000rpm
Top	0.833	28.8
4th	1.000	23.6
3rd	1.396	16.9
2nd	2.087	11.3
1st	2.321	7.1

Final drive gear: Hypoid bevel
Ratio 3.08 to 1

Suspension
Front location	MacPherson struts
springs	Coil
dampers	Telescopic
anti-roll bar	Yes
Rear location	Live axle, torque tube, trailing arms and Watts linkage
springs	Coil
dampers	Telescopic, self-levelling
anti-roll bar	No

Steering
Type	Rack and pinion
Power assistance	Yes
Wheel diameter	15.9 × 15.2in

Brakes
Front	10.15in diameter disc
Rear	9.0in diameter drum
Servo	Vacuum

Wheels
Type	Steel disc; light alloy extra
Rim width	6.0in
Tyres – make	Various; Dunlop Denovo extra
– type	Radial ply
– size	185HR14; 195/70HR14 extra

Rover 3500 (1976–86) – specification *(continued)*

Equipment
Battery	12 volt 68Ah
Alternator	55 amp
Headlamps	Four lamp halogen, 220/120 Watt
Reversing lamp	Standard
Hazard warning	Standard
Electric fuses	12
Screen wipers	two-speed, plus flick-wipe
Screen washer	Electric
Interior heater	Air blending
Interior trim	Nylon velour seats, brushed nylon headlining
Floor covering	Pile carpet
Jack	Screw pillar
Jacking points	two each side
Windscreen	Toughened
Underbody protection	Zinc-coated outer sills, bitumastic overall

Maintenance
Fuel tank	14.8 imp. galls (66 litres)
Cooling system	19.5 pints (inc. heater)
Engine sump	9.5 pints SAE 20W/50
Gearbox	2.8 pints SAE 80EP
Final drive	1.6 pints SAE 90EP
Grease	No points
Valve clearance	Hydraulic tappets
Ignition timing	6 deg BTDC (static)
	6 deg BTDC (stroboscopic at 600 rpm)
Spark plug – type	Champion N12Y
– gap	0.030in
Tyre pressures	F 26; R 26psi (normal driving)
Max payload	1,141lb (519kg)

(Source: *Autocar*)

extra width gave more shoulder room, and there was space for three people instead of just two, in the back of the car. While the new car had an overall width only 3in (76mm) more than the old car, the front track was 5.5in (140mm) and the rear track was 7in (178mm) wider than the old car – this improved roll stiffness and stability. Luggage space and versatility were far better on the new car. In spite of being a much larger car, the new 3500 model was only 23lb (10.4kg) heavier than its predecessor as the result of weight-saving measures by the design team. Finally, the new car's drag coefficient was 0.39, a vast improvement on the much higher-built and more angular P6.

The Rover 3500 was very competitively priced against its rivals (although it was slightly dearer than its main rival, the Ford Granada). It was well-equipped, with standard features including power steering, central door locking and self-levelling suspension, radio, fog lamps and tinted windows. There were servo-assisted brakes, with drums at the rear and discs at the front. The large five-door hatchback design gave buyers virtual estate-car carrying capacity and a very distinctive body style.

The New Rover 3500

Price: Rover 3500 compared

Ford Granada Ghia Auto	£4,329
Renault 30TS	£4,471
Citroen CX 2200 Pallas	£4,535
Rover 3500*	£4,750
Audi 100SE	£4,998
Peugeot 604	£5,306
Opel Commodore GSE	£5,322
Volvo 264 GL	£5,795
Jaguar XJ 3.4	£5,839
BMW 528	£6,471
Mercedes 280 SE	£8,935

* £5,080 as tested, with electric windows, special alloy wheels and 195 tyres, rear seat belts and passenger's door mirror.

(Source: *Motor*)

Performance was very good – the manufacturer claimed a 0–60mph time of 8.6 seconds for the manual car, and 9.0 seconds for the automatic car, with top speeds of 126mph for the manual Rover, and 123mph for the automatic car, better than the performance of most of its rivals. Of course, it also offered the excellent torque of the V8 engine. The SD1 was also economical for a car of this class – the manufacturer claimed that the manual car had an average fuel economy of 26.0mpg and 24.0mpg for the automatic car. The manual car had a fifth gear, with overdrive ratio for economical motoring, a feature which was not available on any of its competitors.

Most of the car's safety features have been described previously. However, Denovo run-flat tyres were available as a factory-fitted option for the new Rover (the P6 had also had this option). The manufacturer, Dunlop, claimed these features:

- extra safety
- control even after deflation
- punctures without tears
- outstanding cornering and road-holding when inflated
- unprecedented run-flat capability
- no spare wheel and tyre required, except in special circumstances
- extensive servicing and repair facilities.

The Denovo tyre was designed to improve safety. In the event of a puncture at high speed, instead of careering out of control as would probably happen with a normal tyre, the car would carry on without any deviations in direction or speed. It was obviously better to be able to continue home, and then change a tyre, than to have to change it, say, on the hard-shoulder of a motorway. (Dunlop claimed that the tyres could be safely driven deflated for up to 100 miles and at up to 50mph.)

However, the Denovo tyres tended to be much better in theory than they were in practice. When the SD1 development team were testing the car in Nevada, they had problems with the Denovo tyres which were fitted, when they carried out tyre testing for 2,000 miles. There were problems with high pressure (caused by the canister inside the Denovo tyre breaking) and chunks of rubber coming off the tyres.

In everyday, practical use, Denovos were not very good in their primary function as

Performance: Rover 3500 compared

	0–60mph (seconds)	Max speed (mph)
Rover 3500	8.9	122.3
Renault 30TS	9.2	114.0
Ford Granada Ghia*	10.5	110.2
Citroen CX 2400	11.6	111.5

* automatic

(Source: *Motor*)

The New Rover 3500

tyres – they wore badly. Eventually, Dunlop withdrew them in the early 1980s after persistent trouble with them and complaints from customers. Dunlop and Rover set up a scheme which allowed Rover car owners with Denovo wheels and tyres to exchange these items for normal ones.

The new Rover had a Triplex Ten-Twenty windscreen. This glass took eight years to develop, and was first produced for Concorde and other advanced aircraft. Its unique sandwich of two sheets of specially treated glass was bonded to a plastic interlayer which cushioned the impact of anyone hitting it and virtually eliminated the risk of serious facial injury in an accident. Like a normal laminated windscreen, Ten-Twenty glass did not craze and impair visibility if struck by a stone.

Public reaction to the car's striking styling was extremely favourable – its high performance and excellent driving qualities were also much appreciated. However, David Bache's specially-styled interior, with its deliberate absence of the traditional comforts of wood and leather, was a different matter altogether. Some people thought the interior looked cheap, with its dull nylon seats and plastic door panels and dashboard. Comments were made about the new Rover's 'Morris Marina interior' – you could almost hear the sound of traditional Rover buyers bursting blood vessels! The skeletal bonnet badge, too, was unpopular with many buyers of the new car, who could soon buy a replacement traditional Rover badge as an aftermarket accessory. Customer demand eventually led to Rover making leather upholstery

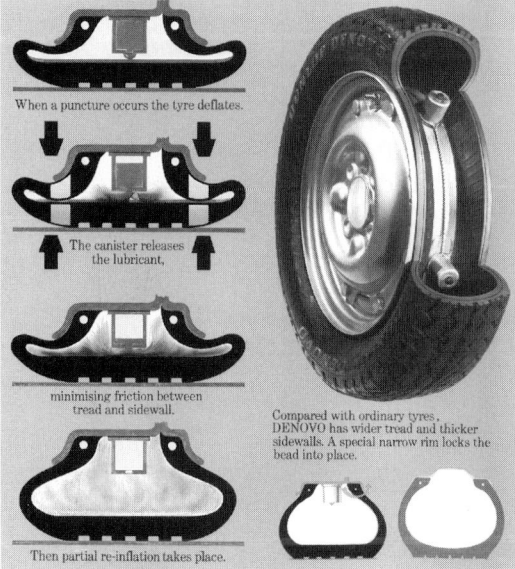

(Above) *The Dunlop Denovo run-flat tyre – an advertising illustration showing how it works.*

The interior was a surprise to those accustomed to being cosseted in an environment of polished timber and leather. The seats, however, were extremely comfortable.

47

The New Rover 3500

The controversial, but distinctive skeletal bonnet badge design (above) *was not liked by many Rover SD1 customers – demand led to the return of a traditional Rover badge* (right) *in 1979.*

Factory fitted options for early Rover 3500

Automatic transmission (three-speed)
Electrically-operated windows on all four doors
Denovo wheels and tyres
Alloy wheels with 195 section tyres
Inertia-reel belts (for rear seats)
Rear view mirror for passenger door

an option in April 1978 and, from late 1979, a traditional Rover bonnet badge replaced the unpopular skeletal design. (Leyland had received previous indication of possible adverse reaction to these items. Twelve months before the car was launched, Rover car owners had been invited to car clinics to look at the new SD1 car. Some of them felt that, although the car was impressive, it was not conservative enough for their tastes – no doubt they would have preferred more wood and leather in the car's interior.)

Early road tests of the new Rover 3500 were very favourable indeed. *Autocar* managed to obtain a top speed of 126mph, with a 0-60mph time of 8.4 seconds, and overall fuel consumption of 20.5mpg for the five-speed, manual version of the car. *Motor*, testing a different car, got slightly lower performance figures of 0-60mph in 8.9 seconds and a top speed of 122.3mph, although they believed that after an owner had put a few miles on the clock, the manufacturer's claimed performance figures would probably be attained. *Motor* also managed to obtain a slightly better fuel consumption figure of 22.5mpg, and gave the car five stars for economy, which was one of the best figures for any car in this class. They believed that driven moderately, an owner would probably be able to get 25mpg from the car, without making any special attempts to save fuel.

The *Motor* testers also gave the new five-speed gearbox a five star rating, commenting that it was one of the best around at the time:

fifth [is] a superb cruising gear giving incredibly low revs at high speed – 3500rpm at 100mph ... Possibly the best feature of all, though, is the change: light, slick, precise, with unbeatable synchromesh – the speed of changes is limited simply by how quickly you can move your hand, and the fourth/fifth dog-leg change, so often awkward or notchy on other five-speeders, is particularly good ... The clutch, with a medium weight and smooth action, complements the rest of the transmission well.

Motor also liked the car's engine:

the Rover V8 provides a steady, continuous build-up of power from idle to maximum without any 'kick-in-the-back' effect ... the general characteristics disguise the fact that invariably you are travelling quicker than you think you are. Moreover, flexibility, in the sense that the engine will pull from seemingly impossible revs, is superb – we were flooring the throttle at something less than 20mph or 700rpm in fifth during our acceleration runs! ... the engine is very well muffled and, with the high fifth gear, is at its best when cruising at speed: at the legal limit it is only turning over at a gentle and to all intents and purposes silent 2450rpm. Even at 100mph it is barely audible.

The car's handling was also awarded five stars by *Motor*:

Another facet of the 3500 that makes it so much of an enthusiast's car ... is the roadholding and handling. The steering is very direct (2.5 turns lock to lock for a 31ft turning circle) and is a delight with an exceptionally quick and precise response. The weighting of the steering is also just right: better than that of the Jaguar or Citroen, and on a par with the (very good) Mercedes system ... On the limit the handling characteristics change progressively from mild understeer to equally mild oversteer which is almost self-correcting, and below the limit the handling is as near neutral as you could wish. Mid-corner bumps which will send other cars lurching across the road are absorbed without any drama, and the levels of poise and stability approach those of the (all-independent) Jaguar and Mercedes and are superior to most other cars whether independently sprung or not ... Roll is only modest, and even strong gusts of wind barely affect the car, so straightline tracking is excellent, thanks no doubt in part to the body shape and front spoiler.

Other aspects of the car were considered good, including ride comfort, which 'for a live-axle car, is generally excellent', the amount of space, especially for luggage, comfortable seats, heating and ventilation system, amount of equipment and instruments. Minor quibbles included the brakes, which were spongy, a dislike of the oval steering wheel, and poor rearward visibility – 'the marked slope of the rear window restricts visibility through it'.

Summing up, *Motor* said:

How glad we are, therefore, to say that the new 3500 is superb and deserving of the highest praise ... Compared to its rivals it is better in so many ways: the Mercedes [280E] may (just) have the edge in performance but it is much thirstier and costs nearly £2000 more ... None of the others [BMW 520, Citroen CX2200, Ford Granada Ghia or Jaguar XJ6 3.4] can match it in performance or economy, few handle as well, and none have such a good gearchange. It is this combination of qualities that makes the Rover the 'best buy' in its class.

Autocar was, if anything, even more enthusiastic about the new Rover 3500 than *Motor* had been. They, too, praised the car's performance, fuel economy, handling and 'colossal

The New Rover 3500

Leyland publicity photographs of the new Rover 3500. It was extremely innovative of the company to introduce a large, hatchback saloon in the executive sector of the market.

The New Rover 3500

The Opposition: Rover 3500 compared

	Engine Capacity (cc)	Output (bhp/rpm)	Gearing (mph/1000)	Tyres (size)	Wheelbase (in)
Rover 3500	3,528	155/5250	28.8	185-14	110.5
BMW 2500	2,494	150/6000	19.8	175-14	106
Jaguar XJ3.4	3,442	161/5000	27.5	205/70-15	113
Lancia Gamma	2,484	140/5400	20.2	185/70-14	105
Mercedes 280SE	2,776	185/6000	19.5	185-14	113
Peugeot 604	2,664	136/5750	19.4	175-14	110
Volvo 264GL	2,664	140/6000	19.8	185/70-14	104

	Length (in)	Width (in)	Empty weight (lb)	Payload (lb)	Tank Capacity (gallons)
Rover 3500	185	69	2,895	1,245	14.8
BMW 2500	185	69	2,998	1,036	17.2
Jaguar XJ3.4	195	70	3,717	750	20.0
Lancia Gamma	180	68	2,910	992	13.5
Mercedes 280SE	195	73.5	3,560	1,145	21.0
Peugeot 604	186	70	3,263	1,071	15.5
Volvo 264GL	193	67	3,195	882	13.3

(Source: *Autocar*)

space provided under the fifth door'. They also liked the steering and the ride, noting:

> The excellent feel sets a new standard for British power steering ... the ride, like the handling is consistent. Allied to the long wheelbase, it means the new Rover rides better than any British car except, perhaps, for Jaguar, Bristol and Rolls-Royce; in its firmness and lack of 'wallow' it is perhaps closest to the Bristol.

In conclusion, they commented:

> It is hard to be over-enthusiastic about the new 3500; on every score, its qualities justify any kind of enthusiasm. It would have been hard to predict, especially looking at the bald paper specification, just how well the car would perform, handle and ride. Add to that the spaciousness and aerodynamic efficiency of the body, and the attention paid to ensuring that the car will last, and it is easy to see why all competitors are casting worried glances, not only at the car but also at its price. If the 3500 can be built in sufficient numbers, if the quality can be maintained along with the price, and if the ground is not cut from under its wheels by ill-advised legislation, the new 3500 should be one of the successes of the decade.

Testing the automatic version, *Autocar* obtained performance figures of a top speed of 119mph and 0–60mph in 10.3 seconds, with overall fuel consumption of 18mpg. *Motor* managed to get slightly better figures of 120mph and 0–60mph in 9.6 seconds, with overall fuel consumption of 20.6mpg. The Borg-Warner automatic transmission was found to change smoothly. However, when *Motor* tested their car in 1977, quality control was not so good – the door sealing was poor (it was possible to see daylight around the nearside rear door, past the seal – which created a loud roar at speeds of over 70mph) and the car suffered from a steering vibration, which may have been caused by poor wheel balancing.

Other journalists were also impressed with the new car, John Bolster, writing in *Autosport*, liked the new Rover, testing both manual and automatic versions. He commented:

> Let us hope that the usual industrial nonsense does not bedevil its future success ... The new Rover is a very important car, which has been designed by engineers rather than salesmen ... That some traditional Rover owners will be shocked there is no doubt, but let me tell them this: as a high-speed luxury car this one knocks spots off the old 3500, and it will use less petrol, too.

Testing a Rover 3500 automatic, *Car* magazine took the car to Germany, and were impressed by its performance, handling and roadholding. They commented that the car:

> never felt lost when faced with the quicker Mercs and BMWs ... there is enough at the top to maintain a nice edge of safety ... nor is it outclassed on twisty minor roads, for it is here as clearly as anywhere else that the Rover reveals that it is a car conceived by engineers who are drivers and developed by drivers who are engineers. You are aware that the steering wheel is servo-assisted, but it is also rack and pinion and is sharp and accurate. The car feels taut and fairly compact and you can scoot it really smartly along tight country roads ... Essentially, you cannot find worthwhile fault with the Rover's handling and its roadholding is almost as impressive ...

Generally, for roadholding and handling the Rover does rather better than many of its German rivals and it is our suspicion that the Rover is a rather more sporting car than many people, including a lot of its owners, suspect. We certainly do not have any reserves about its ability to entertain its driver while conveying him along the road swiftly and safely.

Motor Sport also liked the new, different Rover 3500, and found its five-speed gearbox made it 'a very pleasantly-restful car to drive on long journeys'. The only drawbacks noted were that the instruments were difficult to read, and that 'I would prefer leather upholstery in a Rover'. The writer concluded:

> Overall, I can extend the very highest praise to the Rover 3500, which should be welcomed as just the car Britain needs to stem the Imports race. With the electric windows I would order it costs under £4,850. Compare this with £6,471 charged for the six-cylinder 2.8-litre BMW 528, which would be one personal alternative. Good on you, Solihull, and Great Britain!

Motor Sport were later impressed with an automatic version of the SD1, after driving from Wales to Solihull 'in such quiet luxury' that the editor described the latest Rover as 'the Thrifty-Executive's Silver Shadow'.

Regional newspapers also liked the new car. *The East Hampshire Post*, for example, said:

> It is a hatchback, yet it is in the limousine class; it will cruise effortlessly at high speed and low revs, yet it has the performance of a sports car. The Rover 3500 ... is a car with simple lines, yet its style and grace make it a head-turning event without the distinctive but subtle growl of the V8 engine. And for its

The New Rover 3500

The non-circular steering wheel was not always appreciated by road testers – however, the design team intended that all the instruments could be clearly read, whilst providing a reasonable amount of thigh clearance.

size it boasts remarkable economy ... using this particular car ... provided me with a pleasant experience which ended all too soon ... It is a large car, and one appreciates this as soon as one has to look for a parking spot or has to reverse, but the power-assisted steering and magnificent steering lock – which enables it to turn in a far smaller area than many popular small cars – more than compensate ... such is the smoothness that one sometimes has to watch the speed ... the car requires very little effort to drive.

Generally, the new car was well liked. Minor quibbles included the shape of the steering wheel, the non-traditional interior,

An early Rover 3500, evidently fastidiously maintained by its owner during the late 1990s.

and the ill-sited position of the handbrake (next to the passenger's seat). *Car* magazine made an amusing comment about the favourable media reaction to the new Rover SD1:

> So unqualified was the hymn of praise, led by the unholy sect of advertisement managers of daily papers and some of the specialist press, that the proverbial reader from outer space would have gained the impression that the new Rover 3500, the SD1 of our scoop of months ago, was indeed the only car produced on Planet Earth.

Not surprisingly, given the rave reviews, the new car won a number of motoring accolades. By far the most important of these was being voted Car of the Year in January 1977, by a panel of forty-nine motoring journalists from fifteen different European countries. It was unusual for a British car to win this award, as British designs were not well regarded by Europeans at this time. The Rover won the award by a considerable margin, beating a number of other cars including the Audi 100, Renault 14, Ford Fiesta, Volkswagen Golf Diesel, Lancia Gamma, Mercedes 123 series, Volvo 343, BMW 630 and 633 models, and the Porsche 924. The Finnish judge, who placed the Rover at the top of his list, was quoted as saying 'The Rover proves the British can if they will'.

In May 1977, the new Rover won its second major award, the Don Safety Trophy, justifying the effort put into this aspect of the car's engineering by the design team. Despite a strong entry for this award, the panel were unanimous in their verdict:

> no other car producer for the first time in 1976 appears to have so many safety measures clearly incorporated at the design stage.

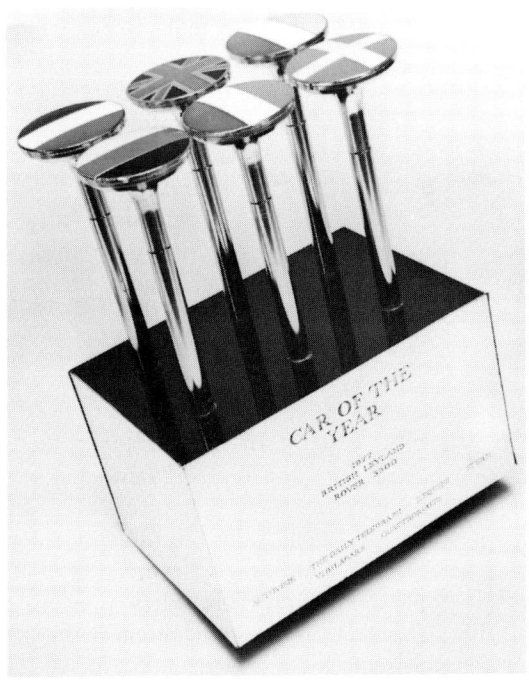

In 1977, the Rover 3500 won the ultimate accolade of being chosen Car of the Year by a panel of forty-nine motoring journalists throughout Europe.

The judges stated that the Rover was ahead of the game in both primary and secondary safety, with its excellent handling, roadholding and fail-safe braking, reinforced passenger cell, deformable ends, optional Dunlop Denovo tyres and Triplex Ten-Twenty windscreen.

In 1978 the Rover was voted Tow Car of the Year by Caravans International. The 3500 was an excellent tow car – the V8 engine had good torque and performance and carefully selected transmission ratios provided optimum tractability. Self-levelling rear suspension, combined with the car's long wheelbase, and wide-track stability, helped to make the Rover a superb tow car. As well as this, it had excellent steering, with good 'feel' letting the driver know exactly what the front wheels were doing, and a good steering lock.

The Rover also won the AA gold medal as 'a major contribution to the safety, comfort, economy, enjoyment and advancement of motoring' (the Rover 2000 having also won this award previously). Rover won the Style Auto award, which was given by a panel of leading European car designers. British Leyland claimed that no British car had ever won so many awards in the first year of its life.

Even as late as 1979, the Rover 3500 was still winning awards. In April of that year, it was chosen as *What Car* magazine's car of the year in the director's car class. Plus points were the versatile hatchback shape, powerful engine and power steering. Minus points included early reliability problems and poor rear three-quarter vision.
What Car said:

> An exceptional car among some very strong opposition, the Rover 3500 wins its place at the head of our rankings for its marvellous all-round competency.

It may be difficult to comprehend the fact now, but such was the demand for the new Rover 3500, and the size of waiting lists that grew for the car, that a sort of 'black market' appeared for these cars, with potential purchasers placing advertisements in car magazines offering to pay far more than the manufacturer's list price.

From the start, the new SD1 was in short supply – a mere 6,816 cars were sold in Britain in 1976. Even in 1977, the first full year of the car's production, only 12,374 cars found buyers. Only in 1978 did orders start to be properly fulfilled. Home sales in that year amounted to around 31,669 cars, about the same number of sales as were achieved by the Rover P6 in 1971.

The car had been shown to 2,000 fleet buyers, and Leyland had committed themselves to supplying more than 500 cars on the first day of the car's launch to fleet buy-

---

Rover 3500 – specification changes 1976–79

1977	Standardization of a passenger door mirror, electric window lifts, and a four-speaker radio/cassette unit (all previously extra-cost options). Single supporting strap for rear parcel shelf was changed for two straps, and the chromed windscreen washer nozzles were replaced by domed plastic items.
1978	Sliding steel sunroof extra-cost option. Leather upholstery available as an extra-cost option (although initially only in one colour – nutmeg brown). Inertia-reel rear seat belts standardized, metallic paint available as no-cost option.
1979	Black paint available as extra-cost option. Borg-Warner type 66 automatic transmission replaced earlier type 65 (although ratios remained as before). Alloy wheels of a slightly different pattern became available for use with Denovo tyres. Ordinary alloys became standard. New rear badges with larger lettering describing the car as a '3500 V8', plus V8 badges on the front wings and a traditional Rover badge on the bonnet to replace the skeletal design. Wheel centre emblems also changed to an outline of the traditional badge. Air conditioning became optional, and there was carpet material on the rear parcel shelf to stop items stowed there from sliding about. A headlamp wash/wipe also became optional, and the power-assisted steering was improved with a new integral pump and reservoir.

The New Rover 3500

ers. One former Leyland car salesman recalled that early in 1977, he had to deal with enquiries from irritated fleet buyers who had ordered SD1s and weren't getting them. The number of orders grew and grew The director for sales and marketing, Keith Hopkins, was quoted as saying:

> Response to this Rover is one of the most impressive I have ever seen for a new car. What we need now is a steady supply of cars with guaranteed quality. Even with an outstanding car like this we cannot afford to keep people waiting too long.

The SD1 was Rover's first experience of using a Ford-style product planning department. Product planners researched who would be likely purchasers of the new car. The product planning department perceived that the main competitors to the new Rover 3500 would be the Ford Granada, the Volvo 244 and the Citroen CX, with lesser rivals being the BMW 520, Audi 100, Mercedes 230, Renault 30 and Peugeot 604. They were certainly correct in identifying the Ford Granada as a serious threat to future Rover sales. The German-built Granada was popular with fleet buyers, and was the major threat to all the various Rover SD1 models throughout the car's production life.

Given the demand for the new Rover, it appeared that in the early days Rover did not need to do a great deal to advertise and promote the new SD1. However, they promoted it vigorously, both in the national press and in motoring magazines and so on. They used the highly unoriginal slogan of 'Tomorrow. Today. The new Rover 3500'. (The slogan 'Tomorrow's Car Today' had been used previously by numerous other motor manufacturers.) Some early advertisements such as those seen in *Country Life* magazine, harked back to earlier Rover advertisements like those used for the Rover P6, and emphasized the traditional

The conventionally styled Ford Granada was perceived by Leyland product planners as the main rival to SD1 sales.

Rover virtues of safety, comfort and excellence. Sales brochures and advertisements placed great emphasis on the new car's safety features. They also noted the value for money, aerodynamic shape, amount of interior space and luggage-carrying capacity, as well as its performance and refinement. Ironically, reliability was also emphasized, as were the number of quality control checks supposed to be made on the new car (to say nothing of the new thermoplastic paint process, which ensured 'a durable quality finish'. Indeed, there may have been a few unfortunate early SD1 owners who would have answered the slogan, 'Tomorrow, wouldn't you rather be in a Rover?' with

This is an early Rover SD1 advertisement from Country Life *magazine, and echoes Rover advertisements from an earlier period.*

Early Rover SD1 paint and trim colours	
Exterior Colours	*Trim*
Richelieu (dark red)	Caviar or Coriander
Pendelican (white)	Nutmeg or Caviar or Amontillado
Midas (metallic gold)	Nutmeg or Caviar
Brazilia (dark brown)	Nutmeg or Coriander or Amontillado
Turmeric (bright yellow)	Caviar or Coriander
Caribbean (metallic bright peacock blue)	Caviar or Coriander or Amontillado
Platinum (metallic silver-grey)	Nutmeg or Caviar

Note: metallic colours used on Rover cars for the first time ever. Trim colours are all different shades of brown.

(Source: Rover 3500 colour and trim card)

'No – not when I'm broken down on the hard-shoulder of the motorway when the electrics have packed up.'

Some people involved in the SD1 project believe that the whole Leyland Group (and former Leyland bus 'super salesman' Lord Stokes in particular) gravely underestimated the overseas competition for all of the British Leyland Group's cars. In 1973, a £500 million expansion programme was intended to boost British Leyland's entire output to 1.5 million vehicles a year, backed up by the Ryder report. This was totally unrealistic. Stokes himself did not understand what foreign competition was all about, and was quoted as saying that 'anyone who buys a foreign car is bonkers!'

The New Rover 3500

Forget the Aston Martin or Lotus – the car James Bond (alias actor Roger Moore) really drove was a Midas Gold Rover 3500. This publicity photograph was taken outside Pinewood Studios at the time of filming The Spy Who Loved Me.

Employees remember that when Michael Edwardes became chairman of the company in 1977, he completely transformed things. Edwardes turned the company round, concentrating on the core business of building cars in a much smaller way than had been envisaged by Ryder. He also carried out a policy of much-needed rationalization and rebuilt the company's model range.

The attitude to the customer which prevailed at British Leyland at this time was hardly conducive to ensuring that the customer would come back to buy another car. The old Rover company had believed that the customer was always right, and was noted for giving good service and spares back-up after they sold their cars. The British Leyland ethos was committed to moving the metal in record time. The customer was looked on as a fool – the aim was to take as much money off him as quickly as possible. Leyland did not have proper respect for the customer.

However, the perceived size of the market for the SD1, which was greatly over-estimated by British Leyland, was compounded by other factors. The economic climate at the time of the car's launch in 1976 was very different from when the SD1 project had been conceived. The mid-1970s oil crisis did nothing for the sales of big cars with large eight- and six-cylinder engines – small cars with good fuel economy were favoured by the public.

Ultimately, far fewer SD1s were sold than originally envisaged by the company, when it had been intended that the SD1 would outsell the combined total numbers of Rover P6 and Triumph 2000/2500 models.

5 The Six-cylinder SD1 Cars – 2300 and 2600

The two six-cylinder versions of the Rover SD1, the 2300 and 2600, were finally launched (after many delays) in October 1977. In practice, though, the 2300 model was not available until spring the following year – in May 1978 it eventually appeared in British Leyland car showrooms all over Britain. The earliest six-cylinder cars had

After much delay the 2600 model was finally launched in October 1977. This photograph shows the cars in suitably grand surroundings.

The Six-cylinder SD1 Cars – 2300 and 2600

	overall length ft in (mm)	overall width ft in (mm)	weight lb (kg)	power bhp/revs	torque lb ft/revs
Six-cylinder Rovers compared with old models					
Rover 2600	15 5 (4,700)	5 9.8 (1,773)	2,866 (1,301)	136/5000	152/3,750
Triumph 2500S	15 2.3 (4,630)	5 6.5 (1,690)	2,696 (1,224)	106/4700	139/3,000
Rover 2300	15 5 (4,700)	5 9.8 (1,773)	2,787 (1,265)	123/5000	134/4,000
Rover 2200 TC	14 11.3 (4,554)	5 6 (1,677)	2,854 (1,296)	115/5000	135/3,000
Triumph 2000	15 2.3 (4,630)	5 6.5 (1,690)	2,590 (1,176)	91/4750	111/3,300

(Source: *Motor*)

'S' suffix registrations. These cars had the same body shells, rear axles and transmissions as the earlier 3500 SD1 cars, but there were differences in their running gear, specifications and small details including badges. Motor journalists noted that these new cars looked virtually identical to the 3500 Rover. Philip Turner, writing in *Motor*, commented:

Outwardly, the new Rover sixes differ so little from the 3500 that you may well have encountered one on the road without knowing it. The only disguise required was to change the 2300 or 2600 script at the rear for that of the 3500. Then only a keen ear would have detected the very different exhaust note. Apart from the wheel trims, the 2300

A well-maintained, early Rover 2300 model in a rarely seen colour – Avocado green.

Rover 2300 (1977–86) – specification

Engine Front, rear drive
Cylinders 6
Main bearings 4
Cooling Water
Fan Viscous
Bore, mm (in) 81mm (3.2")
Stroke, mm (in) 76mm (3.0")
Capacity, cc (cubic in) 2,350cc (92.5 cu in)
Valve gear ohc
Camshaft drive Chain
Compression ratio 9.25 to 1
Octane rating 97 RM
Carburettors 2SU HS6
Max power 123bhp (DIN) at 5,000rpm
Max torque 134lb ft at 4,000rpm

Transmission
Type Five-speed all-synchromesh manual optional (4-speed gearbox standard)
Gear   Ratio   mph/1000rpm
Top   0.83 to 1   24.9
4th   1.00 to 1   20.7
3rd   1.40 to 1   14.9
2nd   2.09 to 1   9.9
1st   3.32 to 1   6.2
Final drive gear Hypoid bevel
Ratio 3.45 to 1

Suspension
Front – location MacPherson struts, lower links
       springs Coil
       dampers Telescopic
       anti-roll bar Yes
Rear – location Live axle, torque tube, Watts linkage
       springs Coil
       dampers Telescopic
       anti-roll bars No

Steering
Type Rack and Pinion
Power assistance Optional
Wheel diameter 15.5in

Brakes
Front 10.1in diameter disc
Rear   9.0in diameter drum
Servo Yes

Wheels
Type Pressed steel disc
Rim width 5.5in J section
Tyres – make Various
      – type Radial
      – size 175HR 14

Rover 2300 (1977–86) – specification *(continued)*

Equipment
Battery	12 volt 50 Ah
Alternator	55amp
Headlamps	4-lamp tungsten 40/45 Watt
Reversing lamp	Standard
Hazard warning	Standard
Electric fuses	10
Screen wipers	Two-speed plus manual intermittant
Screen washer	Electric
Interior heater	Air blending
Interior trim	Cloth seats, cloth headlining
Floor covering	Carpet
Jack	Screw pillar type
Jacking points	4: 2 at front, 2 at rear
Windscreen	Laminated
Underbody protection	Bitumastic and plastisol sill cover

Maintenance
Fuel tank	14.5 imp. galls (65.9 litres)
Cooling system	18.2 pints (inc. heater)
Engine sump	11.2 pints SAE 20/50
Gearbox	2.8 pints SAE 80EP
Final drive	1.6 pints
Grease	no points
Valve clearance	Inlet 0.018in (cold). Exhaust 0.018in (cold)
Contact breaker	0.015in gap
Ignition timing	10 deg BTDC (static)
Spark plug – type	Champion BN9Y 14mm taper seat
– gap	0.024–0.026in
Tyre pressures	F28; R30psi (normal driving)
Max payload	1,180lb (536kg)

(Source: *Autocar*)

and 2600 are almost indistinguishable from the 3500, though the keen-eyed may notice the lack of fog lights on the sixes that are standard on the 3500, and that the 2300 has tungsten in place of halogen bulbs in its headlamps. Nor does the 2600 differ much in either its mechanical specification or its interior trim from its V8 stablemate. But the 2300, designed to bring SD1 motoring within the reach of more owners, does incorporate some changes.

The main difference between the new 2300 and 2600 models and the 3500 model was the brand new in-line six-cylinder engines, which were designed by Triumph and built at Coventry. They were relatively simple power units, with belt-driven single overhead camshafts operating two valves per cylinder via Dolomite Sprint-type valvegear in a light alloy head and four-bearing crankshafts. The main difference between the two new engines was that the 2600 had a longer-

The Six-cylinder SD1 Cars – 2300 and 2600

Rover 2600 (1977–86) – specification

As for Rover 2300, except for:

Engine
Bore, mm (in)   81 (3.189)
Stroke, mm (in)   84 (3.307)
Capacity, cc (cubic in) 2,597 (158.3)
Max power   136bhp (DIN) at 5,000rpm
Max torque   152lb ft at 3,750rpm

Transmission   5-speed manual gearbox standard. (3-speed automatic optional)

Suspension
Rear   Self-levelling telescopic dampers

Equipment
Headlamps   4-lamp halogen 60/55 Watt

Maintenance
Max payload   1,230lb (560kg)

(Source: *Autocar*)

Under-bonnet view of the Rover six-cylinder unit, which is visually identical in the 2300 and 2600 models.

stroke engine (84mm instead of 76mm for the 2300 unit). The 2300 was of 2,350cc in size, while the 2600 was 2,597cc.

There were a few differences between the 2300 and 2600 models, although the 2300 had a simpler specification, in both mechanical and trim terms. It lacked self-levelling rear suspension (having variable rate coil springs and normal telescopic dampers instead), and was also fitted with a four-speed manual gearbox and driver's mirror as standard, lacking a passenger door mirror. The 2300 model had a smaller instrument pod with fewer dials (lacking a tachometer, oil pressure gauge and bulb failure warning lamp, for example). There were minor trim differences – it had slightly simpler seat material, which was more heavily pleated than that of the 2600. It also had a rubber trimmed boot, whereas the 2600 boot was trimmed in carpet. The wheels were slightly different, too, being plainer than the black-accentuated wheel design of the 2600. The 2300 also lacked the option of electric windows, which

The Six-cylinder SD1 Cars – 2300 and 2600

Box-pleat seat trim, velour-trimmed door arm-rests and additional instrumentation on the Rover 2600 (right) differentiate it from the 2300 model (left). The 2600 shown is fitted with optional electric windows, as seen in the switches on the central console.

could be ordered for the 2600. The 2300 Rover was also made cheaper by the absence of a number of lights such as glove-box lamps and door-open lamps.

Factory-fitted options for the early Rover 2600

Power steering
Electric window-lifts
Tinted glass
Inertia reel rear seat belts
Radio/cassette stereo unit
Alloy wheels
Front foglamps
Metallic paint

The six-cylinder cars were not equipped with central locking. A number of optional extras could be ordered for both models, including automatic transmission, power-assisted steering, metallic paint, rear seat belts, Denovo wheels and tyres and so on.

These new cars were very well received by most of the motoring press, especially the 2600 model which, with a top speed of around 120mph, was nearly as fast as the 3500's top speed of 125mph. However, the in-line six-cylinder engine was not as refined as Rover's super-smooth V8 unit. On the other hand, the 2600 was more economical, getting an average overall figure of around 23mpg compared with around 21mpg for the 3500. In addition, the 2600 was much cheaper than the 3500 model –

when it was launched it cost £5,800, £1,000 less than the 3500.

When road tests were carried out on manual versions of the new six-cylinder models, *Autocar* managed to obtain a top speed of 118mph and a 0–60mph time of 10.7 seconds for the 2600, with overall fuel consumption of 22.4mpg. For the 2300 model, *Autocar* obtained the slower top speed of 111mph (in fourth gear) and a 0–60mph time of 11.9 seconds, and a worse overall fuel consumption figure of 22.0mpg.

Autocar commented:

Compared with the manual-gearbox Rover 3500 ... the 2600 is not unexpectedly rather slower. This is not to say it is slow: far from it. Maximum speed drops from 123 to 117mph, which is just what one would expect from a power deficit of 19bhp, or perhaps rather less. Acceleration times are likewise affected all through the range, with the magic 60mph coming up from rest in 10.7 sec instead of the remarkable 8.4 sec achieved by the 3500 ... the lower gears are well spaced out below top ... the ratios feel well-chosen and there is never any feeling of falling into a hole between gears.

In absolute terms, the 2600 is a quick car off the mark despite the first gear, which is on the high side ... The Rover's flexibility is good rather than excellent. It will pull happily from 20mph in top gear, but it would not manage 10mph in fourth (though the 3500 would). Surprisingly perhaps, the overall picture is of a sporting character, with plenty of rewards for generous use of the gearbox.

It would be surprising if the 2600 were as quiet as the 3500, and it turns out not to be. There is a throaty rumble from the engine –

Rover 2300 compared

	Price (£)	Max mph	0–60mph (sec)	Overall (mpg)
Rover 2300	5,645	111	11.9	22.0
Citroen CX 2400	5,428	113	11.8	23.5
Renault 30TS	6,125	117	9.8	18.5
Volvo 244GL	6,231	106	11.4	21.3
Ford Granada 2.3 GL (auto)	5,519	98	14.5	20.6
Audi Avant 1600	5,099	100	12.6	26.9
Peugeot 604SL	6,611	113	9.4	19.6
Lancia Gamma Berlina	7,136	-	-	-

	Engine Capacity (cc)	Power (bhp)	Length (in)	Width (in)
Rover 2300	2,350	123	185.0	69.0
Citroen CX 2400	2,347	115	181.0	68.0
Renault 30TS	2,664	125	178.0	68.0
Volvo 244GL	2,127	123	182.0	67.0
Ford Granada 2.3 GL (auto)	2,293	108	187.5	70.5
Audi Avant 1600	1,558	85	180.5	70.0
Peugeot 604 SL	2,664	136	186.0	69.8
Lancia Gamma Berlina	2,484	140	180.0	68.0

(Source: *Autocar*)

mostly, we suspect, an induction roar, and heard loudest during hard acceleration. Far from disliking the noise, most of us thought it lent the car a pleasant sporting air, not out of keeping with the way the engine actually felt. From about 80mph onwards, the noise could be heard at steady speeds also, but 70mph cruising was quiet enough to permit a low-voiced conversation. Road noise is by no means bad, a constant far-away mutter changing little with speed, while wind noise was much better suppressed than in the first 3500 test car.

... the 2600 showed a clear advantage, with an overall test figure of 22.4mpg against the 20.5mpg of the 3500 ... There seems no reason why gentle drivers should not approach 30mpg in the 2600. As to why it should emerge more economical than the V8, one can only surmise that the six-cylinder engine is more efficient when accelerating. It is also true that the lazy punch of the V8 positively encourages drivers to use more acceleration than is necessary in many situations.

As in their earlier test of the Rover 3500, *Autocar* praised the car's steering and handling, comfortable adjustable front seats, the amount of space (especially in the boot) and its side-window demisters. They appreciated the fact that access to the engine was better than that of the V8-engined Rover, which made the six-cylinder cars easier to work on. *Autocar* also liked the car's ride:

Considering it lacks independent rear suspension, the Rover rides remarkably well. There is much to confirm its designer's contention that the self-levelling system more than makes up ground lost in other ways. The ride may not be as good as the Jaguar's legendary smoothness: the Rover exhibits much more the European 'mainstream' ten-

This is a photograph of a Rover 2600 prepared for display with doors and 'B' post removed.

Rover 2600 compared

	Price (£)	Max mph	0–60mph (sec)	Overall mpg
Rover 2600	5,800	117	10.7	22.4
Citroen CX 2400 Pallas	5,498	113	11.8	23.5
Audi 100LS 5E (auto)	5,599	109	11.8	23.3
BMW 520i	6,099	114	10.5	22.4
Peugeot 604 SL	6,695	113	9.4	19.6
Renault 30TS (auto)	5,834	111	11.7	20.2
Triumph 2500S	5,384	105	10.4	24.8
Volvo 244GL	5,963	106	11.4	21.3
	Engine Capacity (cc)	Power (bhp)	Length (in)	Width (in)
Rover 2600	2,597	136	185.0	69.0
Citroen CX 2400 Pallas	2,347	115	181.0	68.0
Audi 100LS 5E (auto)	2,144	136	184.5	69.5
BMW 520i	1,990	130	181.8	66.5
Peugeot 604 SL	2,664	136	186.0	69.8
Renault 30TS (auto)	2,664	131	178.0	68.0
Triumph 2500S	2,498	106	183.3	67.5
Volvo 244GL	2,127	123	182.0	67.1

(Source: *Autocar*)

dency to slight harshness at low speed, smoothing out above 30mph or so ... At any normal driving speed rough surfaces are covered with astonishing calm, the long (and constant) suspension travel soaking up a lot of punishment, and the seats even more.

Minor criticisms included the shape and grip of the steering wheel, and the visibility for the driver, 'The worst point is at the front, where the sloping nose line vanishes at a point determined by the driver's height, leaving him more or less in doubt how much of it he cannot see.'

Autocar summed up their road test of the 2600 by saying:

Perhaps the fairest summing up of the 2600 is to say that it is as good as we expected it to be, and in many ways better. The SD1 chassis was a known quantity, with many virtues. It remained only to see how the new engine shaped up, and to hope no new awkwardness would be thrown up by the new car/engine combination. Actually, the substitution worked very well. The 2600 is slower than the 3500, to be sure, but it is by no means slow when viewed alongside its competitors. It gains in economy, as one would logically expect. It remains a pleasure to drive, the more so because it has found itself a rather sporting character which could not, surely, have been foreseen. Its job will be to back up the 3500 and widen the SD1 market: it is bound to succeed. All that is needed, as always, is the production and the quality control.

The Six-cylinder SD1 Cars – 2300 and 2600

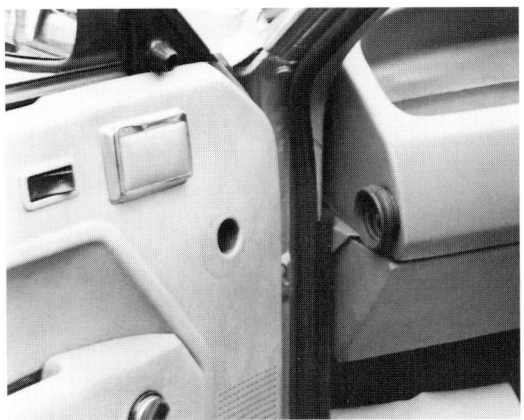

Side-window demisters were standard on all SD1s: air was ducted through a channel right into the structure of the side doors and blown up across the front side-windows, keeping them clear at all times.

By the time of the launch of the six-cylinder models, some motoring writers had started to become more critical of the SD1 cars. Roger Bell, writing in *Motor*, commented that 'the new Rover, in whatever guise, does not ride particularly well'. Road tests of the 2600 carried out by *Motor* and *Autosport*, commented on the car's noise when accelerating at high revs, and its lack of good low-speed torque, John Bolster of *Autosport* saying: 'It is in those areas particularly that the new car needs further development, before it can be said to have reached its full potential.' He also remarked:

> The roadholding inspires great confidence and the car feels safe on wet roads. I found it very difficult to judge the handling because of that weird abortion of a steering wheel if you can call something a wheel that's anything but round. I was ashamed to find that a thing so trivial could annoy me so much ... I trust that the present tortured shape will soon go the way of the one on the Allegro.

Other motoring journalists, however, were less critical of the new six-cylinder cars. One did not want to return the 2600 model he had been driving for three days. Another, testing a 2600 automatic, was impressed by its good overall fuel consumption figure of 26mpg (which included gas-guzzling spells in London rush-hour traffic, and other city driving). He was also impressed by the car's vast load carrying capacity, using it to collect a large sideboard with the rear seat folded flat, commenting: 'It is certainly a luxurious way of entering the removals business!' Another journalist was impressed by the appearance of the Rover 2300, stating:

> On first seeing the car I couldn't help but be impressed, for the shape is unique and distinctive, and will most certainly last as well as, if not longer, than the model it replaces. It has that certain touch of class that naturally attracts attention wherever it goes, because people enjoy looking at quality in any shape or form.

When *Autocar* tested the Rover 2300 model a few months later in June 1978, they preferred the 2600 model out of the two six-cylinder cars. They concluded:

> The Rover 2300 completes the SD1 series and has enough qualities to justify its manufacture. However, disregarding the less important options fitted to our test car but including power steering and five-speed transmission, its price is still over £6,000 and within £200 of a similarly equipped 2600, which has much better instrumentation and self-levelling rear suspension. Bearing this in mind, and the lack of any practical economy gains from the smaller engine, the 2600 would seem a far better buy, and is bound to be more sought after on the second-hand market. Although the 2300 has nothing remarkable in the way of mid-range perfor-

mance (prospective towing owners please note) it does offer exceptionally quiet cruising, and the expected excellent SD1 steering, handling, and ride qualities.

Several road tests of the 2300 model commented on the car's lack of low-speed torque, and its inferior ride and fuel economy when compared with the 2600. On the other hand, road testers (for *Motor Sport* and *Autocar*) commented that the 2300 engine was a much smoother and quieter unit than that found in the 2600. *Motor* was even pleasantly surprised by the 2300 model in standard form (without power steering):

> I expected the worst of the cheapest 2300, not because it was down on power but up on steering effort with 4.5 turns of the wheel from lock to lock instead of the 2.75 of the light and super-responsive power-assisted set-up. I'm glad to say it was nothing like as bad as I feared. True, lots more wheel twirling and quite a bit more effort was needed on sharp corners, but only when parking, reversing or U-turning did it become really tedious and cumbersome, and even then no more so than the steering of, say, the Audi 100. Low gearing makes it light enough for normal driving though the extra twirling accentuates the odd shape of the wheel: the flat bit at the bottom, which is meant to improve thigh room, is usually elsewhere.

Motor Sport liked the new car:

> We plumped for a 2300, four-speed manual, power-steered car first, expecting it to be a disappointment after our familiarity with the 3500. Not a bit of it! The engine responds to the key with a crisp, quite sporting note and it loves to rev. The gearbox needs stirring to obtain the best from the engine, but with such a delightful box,
> who cares! Engine noise is not what one would call inconspicuous, but it is not overbearing and is balanced by low wind and road noise. Acceleration lacks the scorching athleticism of the 3500, 0–60mph taking about 11.5 seconds, but once this smooth six is wound up it flies along very rapidly indeed, an easy 90 to 100mph cruiser, even without the optional fifth gear. Leyland quote the maximum speed as 114mph – we saw rather more than that, quite exceptional for such a big car of modest capacity.

Few changes were made to the six-cylinder cars between their introduction and 1980. Leather upholstery was an extra-cost option from April 1978, although very few of these cars in practice actually appeared with this option. Power steering was fitted as standard to the 2600 cars in October 1978, and from February 1979 black paint could be ordered as an extra-cost option. In November 1979, the new 1980 season six-cylinder cars appeared with the new, traditional Rover bonnet badge. There were different wheel trims for both models, and larger lettering on the rear badges. The automatic transmission changed from type 65 to type 66. Air-conditioning became an option for the 2600 model.

By the time of the six-cylinder Rover cars, the fleet car market had become extremely important for the British Leyland Group, and was particularly targeted by the company. In October 1977, about 150,000 executive cars were purchased in Britain. The new six-cylinder cars plugged the gap between the Princess 2200 and Rover 3500 car models, and were intended to capture an even bigger share of the vital fleet sales market.

Motor Sport noted:

> Leyland are tilting particularly at Granada sales with these new models. Having driven both Ford (the new model) and Leyland offerings we can say that they are as alike

The Six-cylinder SD1 Cars – 2300 and 2600

as chalk and cheese. The Rover is definitely more sporting, tauter in its ride and harder on the ears, reminiscent of a BMW 'six'. The new Granada is a 'softer', quieter car. The choice is one of matching human temperament to motor car temperament.

In the ideal scenario for Leyland Car product planners, the managing and other directors of a company would have the premium Rover, the 3500 model, with managers having one of the six-cylinder Rovers, and with staff placed lower down in the company hierarchy (such as salesmen and so on) having the smaller Leyland cars. The Rover 3500 model of the SD1 established itself early on as a fleet car with more than 85 per cent of cars built penetrating the executive fleet sector. Rover SD1 car advertisements even boasted about this. I can recollect seeing numbers of Rover SD1 cars in company car parks. The 3500 model was commonly used by directors.

With regard to selling the new six-cylinder Rover SD1 cars, they replaced both the Rover 2000 and the Triumph 2000 range of cars – an extremely difficult act to follow. Both of these cars had sold very well – even late in their production lives, the combined sales of Triumph 2000/2500s and Rover 2200s in the first five months of 1977 totalled 5,680 cars. Leyland Cars set itself the objective of exceeding the sales performance of both these models with the new Rover 2300 and 2600 cars. Sadly, these aims were never realized. A number of factors (which will be looked at later on) affected the sale of these cars. They were especially badly affected by a slump in the sale of all large cars which followed a big fuel price increase in 1979. As a result of this, in the summer of the following year, the six-cylinder SD1 cars were subjected to considerable price discounts. This helped Leyland car dealers to clear out their old stocks of

The six-cylinder SD1 cars were particularly aimed at the growing fleet sector – the company aimed to take some sales away from Ford's Granada.

cars in time for the new upgraded models in 1981.

Nigel Heslop, who was involved with product planning for the SD1 range of cars, can remember that the company perceived that the 2300 model would be the biggest success in sales terms of all the SD1 models. This proved to be completely incorrect, in fact, only about 43,000 2300s produced.

With its brand new (Triumph-designed) engine and the fact that it was the most popular single model in sales terms with 108,572 produced altogether – although more V8-engined versions of the car including Vitesse and V8-S models were manufactured in total – the 2600 model was, perhaps, the definitive version of the Rover SD1.

6 Wheel of Misfortune – Making the SD1

It is impossible to relate the Rover SD1 story without examining the problems which affected the car's manufacture. Problems in the British Leyland company as a whole (including a number of industrial problems with the work force) had a dramatic and unfortunate effect on the quality of the new SD1 cars when they first came out. Some of these problems took a long time to put right.

The whole Rover SD1 project cost the British Leyland organization some £95 million – this included the costs of designing and developing the car, a new factory to build the cars (which cost £31 million) and a car body paintshop (which cost £6.2 million). Leyland Cars could not afford to make any mistakes when it manufactured the new SD1, but unfortunately, problems were only too apparent.

The car, which received accolades and good reviews when it was launched in 1976, appeared to have had an auspicious start. Unfortunately, this turned out not to be the case. Soon after the 3500 and six-cylinder engined cars were introduced, horror stories started to circulate about numerous problems with the new car. Body and paint problems on early cars led to corrosion – the bonnet, front and rear wheel arches and lower edge of the hatchback door were particularly badly affected.

The early examples were poorly finished and badly built – this would have been an unforgivable shortcoming on the humble Mini, let alone the Rover SD1, which was intended to be British Leyland's flagship. There were electrical problems, especially with the electric windows and central locking system. (One unfortunate person had to squeeze through the sunroof of his car when the central locking and electric windows packed up, in the middle of a busy high street – such stories did nothing to help the reputation of the car.) The ends of the instrument binnacle tended to fall off. When the six-cylinder cars were introduced, these were plagued by similar unreliability problems to previous Triumph cars – they suffered from oil leaks, gasket failure, consumed excessive amounts of oil, and burned out pistons. Not surprisingly, the car quickly gained a poor reputation with customers. These factors also had a profound effect on the second-hand values of the new SD1 cars – both the 3500 and six-cylinder models depreciated rapidly.

Some of the problems with the new car eventually became apparent in long-term road-tests by motor journalists. For example, *Autocar* magazine bought an early automatic Rover 3500, and ran it for a year and 11,900 miles. Their comments were typical, and illustrate the lack of quality control on early Rover SD1 cars:

> The most disappointing feature about the Car Of The Year was the sad lack of quality control during building and the minimal pre-delivery inspection. Most major fault was a gap between windscreen and pillars,

Solihull – the new Rover car production plant

The new Rover production plant which was built at Solihull, at a cost of £31 million, represented the biggest single development undertaken by the British motor industry for forty years. The site was chosen for its good road communications, its central position, proximity to suppliers and the availability of motor industry-trained staff in the West Midlands.

The scale of the project was massive, and construction work began in July 1973. Within the 100-acre site there was a 23-acre single-storey assembly hall and a 4-acre three-tier paint plant. The factory itself covered an area of 1,500,000 square feet. Total conveyor length in the assembly hall was 1.25 miles. The new plant employed 3,200 operative staff and had a potential operating capacity of 240 new Rovers per day.

Unpainted car bodies arrived at the plant from Castle Bromwich six at a time in double-deck transporters, to be fed into a ground-floor storage and marshalling area. In the paintshop they were cleaned, primed, underbody-protected and colour-sprayed by an automatic system in a dust-free environment. The new thermoplastic paint process involved automatic spraying and high-temperature oven-bake, all the ovens being housed on the top floor of the plant.

After leaving the 900ft-long paint plant, the bodies travelled along a bridge link to the assembly hall, where interior trim was installed on the three 1,400ft-long conveyors. Engines and gearboxes were prepared at the end of the hall, and the trimmed bodies were moved to three mount lines, each 800ft long, for the addition of mechanical components.

Extensive provision was made at the end of the tracks for road test rigs, pollution testing, inspection, rectification, water testing, final assembly, valeting and protective waxing.

From April 1980, TR7 and TR8 cars were assembled alongside Rover SD1s, until they ceased production in October 1981.

In spring 1982, as part of a process of company rationalization and factory closures, the Solihull factory was temporarily disused, and SD1 production moved to Cowley. Eventually, major components for Rover's 4 × 4 vehicles were produced on the SD1 plant.

(Source: Leyland Cars)

The new Rover SD1 production plant at Solihull was Europe's most modern car assembly and paint facility in 1976; it covered over 1,500,000 square feet. In common with the rest of the British Leyland empire, it was not without difficult labour relations, in addition to the technical problems encountered.

On the production line, after the bodies have just been painted.

which allowed in rain and draughts. Hatchback door was badly fitted, and the front doors were rehung and adjusted to get them to close properly and to cut down wind noise. The general fit and finish was also poor.

In spite of these faults, the writer still apparently liked the car, noting that:

> Once these troubles had been sorted out, the Rover came into its own, with good performance carried out in a smooth, quiet manner. The compact size of the car, and the ability to turn it into a near-estate car load carrier makes it truly versatile ... At the moment the car has developed a slight rattle from the engine, and the driveline produces some odd clunks when moving off. Apart from this, and the initial faults, the Rover has been a superb car.

Car magazine also carried out a long-term test on a 1977 automatic Rover 3500, running it for over 20,000 miles. They reported similar problems. The car was delivered with a badly leaking steering rack and servo unit filter – this had to be replaced twice. There was a breakdown at 6,000 miles caused by a broken needle in one of the SU carburettors, which required urgent roadside repair (this typical fault was eventually rectified by SU). The fusebox lid frequently fell off, and the car suffered from excessive, uncured wind-noise. The metallic paint was too easily chipped. Other things which had gone wrong were less serious, such as a blown headlamp bulb. The car's interior and trim details came in for particular comment:

> The finish in the boot annoys us too; it is carpeted, but it looks more like a DIY job than something stemming from Britain's most modern car factory. There has been a lot in various correspondence columns about the Rover SD1's detail finish: our experience is that the odd bit does fall off, and that occasionally things do disintegrate. The latest bit to go on our car is the plastic cowling under the driver's seat which just fell to bits ...
>
> It needs (and deserves) to have silly things like the windnoise eliminated, it should have a more appealing dashboard and better instrumentation, the feeble

plastic bits should be replaced by good quality fittings and the cabin would benefit from more attractive upholstery ... Living in London, it also becomes apparent that the bumpers are too frail to fend off people who park by ear, and while most cars in its class now have some sort of protection along their flanks the Rover has none.

You would think from all these comments that *Car* magazine hated the Rover SD1 car – you could not be more wrong. The writer commented on the:

> Generally satisfying experience with a car that has fulfilled the expectations we held for it. Smooth and easy to handle around town; fast, effortless and safe on the motorway. Strong performance with good fuel consumption considering engine size and the presence of automatic transmission. Ride generally good ... some compensation through self-levelling facility and always excellent handling coupled with good roadholding ... Low running costs and reliability vindicated designers' choice of straightforward mechanicals. Overall, very impressive big saloon with perhaps more to offer keen drivers than those merely seeking prestige.

Other writers were irritated by other features of the SD1 cars. One journalist in *Autocar* disliked 'the sadly cheap-sounding clang with which the doors shut – most inappropriate for a car of this class.'

Early six-cylinder engined cars fared no better than their V8-engined relatives. Some owners of these cars experienced a number of problems. One of these was heavy oil consumption, caused by seepage of oil down the valve guides into the combustion chambers. These engines had a tendency to become noisy if the valve clearances drifted apart – resetting these clearances could be both time-consuming and expensive. Other common problems included some piston burning, failed head gaskets and serious oil leaks.

Philip Turner, writing in *Motor*, carried out a long-term test of a Rover 2600, running the car for over 50,000 miles and keeping it for more than two years. Early on, he commented:

> When people ask me what car I'm running and I tell them 'A Rover 2600', their reaction is to demand almost aggressively, 'Well, what has gone wrong so far, what has fallen off?' ... after the first Running Report on the car was published, I had several letters from readers saying how lucky I was that my particular 2600 was so reliable, for the examples they owned seemed to have given nothing but trouble ... I find the more I drive my Rover the more I like it.

Problems affecting Philip Turner's car included: the boot trim coming adrift, water leaking into the boot (turning the spare wheel well into a swimming pool). There were problems with handbrake cables – these tended to seize in the 'on' position (though never enough to lock the brakes on so that driving was impossible), and had to be either attended to or replaced. Other items to give trouble included the water pump, clutch master cylinder and power steering pump leaking – and head gasket failure (to name just a few). The car's paint had not proved to be very durable, either – after just 50,000 miles, rust was becoming evident ahead of the rear wheel arches and on the rear quarter above the right rear wheel arch.

Autocar, too, carried out a long-term road test of a Rover 2600 automatic, experiencing similar problems to Philip Turner. They called it 'Friday's child' (a car which gave trouble from new, so the story went, was built on either Monday or Friday). The back axle of the car hummed from new, and had to be replaced at 4,000 miles. It burned oil,

Wheel of Misfortune – Making the SD1

Workers assembling initial body fittings and other parts to the newly painted shells.

Lowering the V8 engine into position – a stage in the assembly of the running gear of the car.

so the valve seats had to be replaced by new ones. A window-winder fell off and had to be replaced. Other problems included a leaking petrol tank, and water pump and power steering pump failures. The heated rear window element had to be repaired (due to a bad earth) and a new connector was fitted to the heater motor. Other comments included:

> After 2½ years the Rover 2600 Automatic leaves us with mixed feelings. Little things saddened me; the way the facia and instrument binnacle covering PVC material is crudely creased and stuck down at corners, the doors shut with a certain tinnyness not found on cars that cost half as much, also, the self-propping bonnet stay is awkwardly positioned on the passenger's side of the engine bay so one has to run round the front of the car to release it ... without a partial respray the bodywork would now be very tatty ... Rattling noises from the hatchback area indicate a degree of poor breeding in a car of such good looks and distinguished pedigree ... The Rover 2600 is a popular candidate for company purchase for middle and senior management. Though it stands up well in specification and performance with foreign competition from manufacturers like Audi, Mercedes and Citroen, ours has fallen down in reliability. This is a mantle of notoriety that it cannot afford to keep.

Nevertheless, *Autocar* liked the Rover 2600, stating:

> Our Rover is usually a joy to ride ... smooth, quiet, lively, with acceleration answering every call. It starts well and passengers get a comfortable ride ... In spite of all the problems I grew quite fond of it.

Wheel of Misfortune – Making the SD1

The problems described previously were only too typical of those experienced by owners of early Rover SD1 cars. Many owners had ambiguous feelings, torn between love of the car, and their dislike of the way it was put together, fell to bits or broke down. Some felt like the *Car* magazine journalist, who remarked 'Still, we wouldn't hesitate to buy another one and that's what counts'. Many SD1 owners returned to buy another one, despite all of the faults.

Why was it that the Rover SD1, which had been conceived as the saviour of British Leyland, proved such a disaster? What exactly had gone wrong?

Industrial problems, both within Leyland and at their suppliers, did nothing to help matters. When the SD1 was launched onto the European market at the 1977 Geneva Motor Show, this was severely disrupted by a shortage of cars to sell, caused by a dispute at the Castle Bromwich body plant and a toolmaker's strike. Leyland only managed to get a few cars from dealer demonstrator stock to put on display – there were no cars available for dealers to sell to any eager Europeans who wanted to actually buy the new car. At home and abroad, Leyland dealers could not quote actual delivery dates.

At the new plant, built just to make the new model Rover SD1, industrial relations did not go well. Here, there was a combination of demarcation disputes and overmanning due to the end of piecework. People involved with the project can remember that there were many problems with the unions over job demarcation – for example, if a car had a minor problem, such as a reversing light which failed to work, several different people were called on to sort it

The body drop, where the fully trimmed body is lowered onto the car's running gear. The Watt's linkage for the lateral location of the rear axle is clearly visible in the foreground.

Solihull – the new Rover paint plant

An important part of the new Solihull plant was the car body paintshop, built to paint the new Rover SD1 cars. It cost £6.2 million and covered four acres. The paintshop had a floor area of 900ft x 156ft and a total conveyor length of 4.1 miles.

In all, thirteen processes were involved in the new paintshop, which was a three-storey building. The plant received the bodies-in-white for storage at ground level. Pre-treatment, body preparation and paint booths were on the first floor. The top floor housed stoving ovens, fresh air replacement plant and fume-eliminating incineration equipment. Fast-moving lifts connected the line conveyors between floors.

Technical refinements to improve body painting included:

- a pre-treatment line with double chain conveyors set wider than the body to prevent spotting;
- electrophoretic painting systems capable of operating with different paint formulations;
- 'Hydrospin' paint spraybooths, to give the best paint booth working environment with maximum cleanliness and efficiency;
- air filtration system which completely eliminated dust in the air supply to the spray booths;
- minimum paint wastage from Drysys automatic painting equipment to give high-speed spraying and fast colour changes;
- power washing prior to colour coat;
- 'Effluair' modern, economical oven heating and air purification system which reused heat from the exhaust gases.

Paint system

Ground floor	Two-tier system of twelve parallel main conveyor lines extending 200ft, six lines for incoming bodies-in-white and six for finishing body storage.
First floor	
1st line	Pre-treatment, seven-stage alkali cleaning and phosphate applications. Force drying of body before total immersion for two minutes in Electrodip primer.
2nd line	Dust and water sealing, underbody protection, hot air blow unit to set off sealer. Tac rag enclosure and through dust-proof tunnel to surfacer application booths (manual and automatic).
3rd line	300 feet wet sand deck with 13 rotary sanders, high-pressure power wash, demineralized water spray rinse and dry off at 180 degrees F. Flow stick sealer applied, tac rag wipe and spot surfacing.
4th line	Three-line body storage area, damp scuff area and tac rag wipe. A 210-ft Hydrospin booth with four automatic spraying machines to apply final colour coats. Colour variations supplied by ten colour changeover valves. Paint system initiated in mix house alongside main paintshop building. The colour coat was thermoplastic acrylic, the four coats being applied wet on wet to give a 0.0025-in thick coating.
5th line	Body inspection, tac rag wipe and spot colour spray.
6th line	Body inspection, oil sanding and cleaning, black out spraying of styling features.

A further inspection preceded cleaning and reflow tac rag booth. All bodies then passed through a 630-ft hot air reflow oven and were cooled and polished before final inspection.

> **Solihull – the new Rover paint plant** *(continued)*
>
> Top floor
> 1. 420-ft Electrodip bake oven stoved for 15 minutes at 360° F.
> 2. 500-ft oven stoved surfacer for 20 minutes at 320° F.
> 3. Spot surfacer stoving.
> 4. 400-ft colour oven achieved part one of the coating at 180° F for 15 minutes. Part cure only was required at this stage; the mirror finish was achieved by reflowing in later stages.
> 5. Paint re-work stoving in a 270-ft oven for 10 minutes at 180° F and air-cooled.
> 6. 630-ft reflow oven where body temperature was raised to 310° F for a minimum of 25 minutes.
>
> (Source: Leyland Cars)

out, and this could all take several days to put right. Shop stewards kept having people moved around to different jobs which they could not do. Workers commented that management continually speeded up the production lines in order to increase the number of SD1 cars produced. The new plant became a battleground between the old Rover management and its new Leyland masters. Jean Rivers was shop steward on the final lines, and recalled:

> The facilities were much better, because it was a much more modern plant than the old Rover. But facilities don't make a place happy to work in. The first thing the shopfloor always notices is the times they change management. They were changing them every three months. It just wasn't running right.

The Rover SD1 was put together in an unhappy working atmosphere; it is not surprising that it turned out to be so poorly-built.

The electrical problems on the new car were the fault of the suppliers, Joe Lucas 'Prince of Darkness' – the traditional scourge of British cars. Lucas, of course, had their own industrial disputes at this time. However, one of the people involved with the project believes that the SD1 is prone to electrical problems because Lucas were responding to pressure from Leyland to cut costs – also people on the line at Lucas were not putting things together properly.

The new Rover paint plant was one reason for problems. The paint on early cars was prone to chipping and peeling off, especially on the car's bonnet. The problem here was that the new technology that was being used was not adequately proven. The thermoplastic process had been used previously in the USA, but was totally new to British Leyland. There were chemical problems with the paint process on early cars – the final paint coat did not adhere well to the primer. Some colours, such as yellow, were worse than others, such as gold and silver. Some people believe that attempts by production managers to speed everything up also caused problems. (For example, during the early stages of the painting process, the bodies went into a washing bath before phosphates were applied. The washing bath was not changed every 20 minutes or so, as it should have been, owing to managers trying to speed up the process – this would have contributed to early cars corroding

Wheel of Misfortune – Making the SD1

A late Vitesse is seen outside the Rover plant at Swindon, where its body panels were manufactured.

earlier than they should have done.) Ironically, Rover sales brochures boasted that:

> A new thermoplastic paint process gives the Rover excellent colour-retention properties. In addition to the four high gloss top coats each car undergoes four separate preparation and priming processes which ensure a durable quality finish.

Another reason for problems with the new car was the attitude within British Leyland itself, which extended to the highest level of the organization. Peter Grant, the production manager at Solihull at this time, remembered:

> I was at a dance at the Civic Centre in Solihull and a senior director of British Leyland came up to me and said, 'You Rover people are all the same. You worry about quality. We want quantity. We've got to get this SD1 turned out in quantity' ... Morale was very, very bad. We had sensible middle-aged people. They didn't want to be sworn at or screamed at and threatened with the sack if they didn't decide this that or the other. The plant director was despairing of the quality of the cars that were going into sales. It had to be seen to be believed.

Consequently, inspectors were overruled by managers, and sub-standard cars were produced in such quantities that they filled up the factory, and had to be sent on to dealers to sort them out and put them right before they were sold. Rover workers can remember that quality control (or lack of it) was a major problem. There was no effort by the company to improve the situation. In the end, this affected sales. Efforts to cut costs and increase volumes were all-important. This went against the old Rover tradition of putting quality above anything else.

Ray Horrocks, at one point the Chairman of BL Cars, believed that:

> When the SD1 was designed back in the early 1970s there were not enough produc-

tion engineers alongside the design engineers. It was built by engineers for engineers, so the car was productionized as it went down the line. There wasn't enough development done of the car which reflected the lack of testing facilities then available to the company. When you launch a new model, there are a number of things you aim to avoid. They are putting it into production at a new plant, with a new paintshop and with a new engine and transmission. The SD1, particularly in its six-cylinder form, suffered from all these shortcomings.

The people involved in the project can remember that morale was really low in the company at the time, and this led to problems, with people leaving to go and work elsewhere. We have to remember what British Leyland represented at this time – it was the company which introduced us to the delights of the Austin Allegro, Morris Marina, Austin Maxi and Austin Princess. The whole Leyland organization was continually the subject of poor publicity and the butt of jokes. At one point, there were as many jokes on *The Two Ronnies* BBC TV comedy show about British Leyland as there were about British Rail! Needless to say, the people who worked for the company became somewhat tired of this, to say the least. Many people who worked for the company feel that the problems were caused equally by both sides, management and workforce. However, the problems of British Leyland seemed to be encapsulated in a famous photograph taken in 1979, which showed Derek Robinson ('Red Robbo' to the press), the senior shop steward at Longbridge, at a demonstration, in front of a poster showing one of the company's least successful cars, the Austin Allegro (a car which 'not only turned out to be a lemon but also looked like one').

Production targets for the Rover SD1 cars were never met. During the early 1970s it was proposed that 1,500 cars should be produced a week. However, John Barber was keen to increase volume production, and he persuaded the British Leyland Board to double this target to 3,000 cars per week (150,000 a year!) Far fewer cars were eventually produced (and sold) than intended, and the production targets never looked realistic. There was a general slump in the sales of all large cars following the big increase in the price of fuel in 1979. (Curiously enough, six-cylinder Rover SD1 sales were far more badly affected than the V8 cars.) From 1979 onwards, stories circulated about the fields full of Rover cars which could not find buyers. These cars were left out in all weathers for many months at a time. It is not surprising, then, in spite of all the special paint treatment and anti-corrosion measures (including underbody sealant, zinc-coated sills and body sills injected with wax), that many people who actually bought these vehicles found that they went rusty very quickly.

Before production of the car was moved to Cowley, there was a transitional period from late 1981 to early 1982, when new Series II cars were being assembled, and old revised models were also being built at the Solihull and Cowley plants. In the spring of 1982, production of the Rover SD1 was finally completely transferred from Solihull to Cowley, in the Michael Edwardes era of rationalization.

The Rover cars were assembled in Cowley alongside Triumph Acclaim, Austin Princess and Austin Maxi models. Most people believe that the Cowley-assembled cars were of a much better build standard than the earlier Solihull-built ones. The pride of Solihull had to be built at the old Morris factory before its quality improved – Lord Nuffield would have been pleased!

Looking back with hindsight on the various problems which affected the Rover SD1, one former director at the time recalled:

It was a nightmare of a debut for a car which had received so many rave reviews and for which we had so many high hopes. It was two years before it recovered.

It was all too late for British Leyland. By now, the previously untarnished name of Rover had received a major setback.

SD2 – The car that never was

One casualty of the unfortunate British Leyland era was the SD2 project. This was a Triumph project, intended to replace the Triumph Dolomite range of cars in 1979. The SD2 project was seen as being some sort of compensation for Triumph's Puma design losing the competition with Rover for the design that eventually became SD1.

Originally, several designs were submitted, by David Bache, and the Michelotti and Pininfarina design houses. Rover people such as Spen King wanted to use the attractive Pininfarina design – however, they were overruled by directors Stokes and Barber, who chose the Bache design. According to Spen King, the David Bache design had an over-bodied appearance with little wheels. The SD2 was a five-door hatchback rear-wheel drive car, which had decidedly Gallic overtones in the styling of its rear quarters. Prototypes were built, and ran well using basically the same engine, suspension and running gear as the Triumph TR7. Spen King's design team put a great deal of effort into engineering this project.

However, the project was killed off at an advanced stage in the mid-1970s, for a number of different reasons. The car's styling was not believed to look as good as that of the SD1. Leyland was suffering from a severe shortage of cash at this time, and did not have the money to fund both new Triumph and Austin models. Also, it was believed that the car would overlap with the Marina and its replacement, the Maestro, so the company carried on selling the Dolomite for a few more years. SD2 was the victim of company rationalization.

This is a photograph of an SD2 full-size mock-up in the sales showroom in 1973, showing its rather inept styling. If the car had been any later, one could be forgiven for thinking it was the result of a nasty accident between a Lada Samara and a Renault 16 that had found a set of Dolomite Sprint wheels!

7 The Rover V8-S – the Luxurious SD1

At this point in 1979, this photograph for the announcement of the Rover V8-S could be said to be an all-Leyland production, as besides the Daimler, the background was largely filled by an AEC Routemaster bus.

From the early 1970s, the Rover SD1 design team had been working on an improved version of the car, with the aim of launching the car onto the US market. They eventually did this in 1980, but many of these changes were previewed by the arrival of the V8-S model in June 1979. The new model was part of a revision of the whole range of SD1 cars for the 1980 season. It was available only in the UK at first, although the compa-

The Rover V8-S – the Luxurious SD1

ny intended to introduced the V8-S model to other markets later. The V8-S model was the top model of the range of SD1 cars. Sales brochures proclaimed: 'The Rover V8-S. The ultimate difference'.

The V8-S was important for two reasons. First of all, it was the first attempt by the company to respond to customer demand for a better specified car with a more luxurious interior that was more in keeping with Rovers of old. Secondly, it was a rare model – one of the rarest versions of the SD1. Launch stock was 900 cars, but the original production target of 140 cars per week was probably never reached. It is not known exactly how many of these cars were made – 1,040 were produced in 1979, but the figures for 1980 were included with the rest of the 3500 SD1 cars.

Jaguar Rover Triumph (the company name at this point) hoped that the V8-S model would plug a gap in their range. As *Car* magazine commented:

> The middle management pecking order was well covered by the three existing SD1s – 2300, 2600 and 3500 – leaving Jaguar to cater for the top brass, starting with the XJ3.4. But between the most expensive Rover at about £8,500 and the cheapest Jaguar at just under £12,600 there was a breach that BMW, Mercedes and Opel, among others, have not been slow to exploit. Enter the 3500s, at £10,699 the most expensive Rover yet, to reinforce JRT's defences ...

Jeff Herbert, managing director of Rover Triumph Cars, was quoted as saying:

> Rovers are already well endowed with many standard refinements, especially the 3500, but we have taken the V8-S a significant step further up the ladder in terms of specification.

> Competition for sales in the 'upper echelon' of the executive car market is tough and we are convinced that the Rover V8-S will appeal to top business and professional men who look for distinction, a high level of standard equipment and good fuel economy.

Improved interior trim and additional special equipment set the V8-S apart from previous Rover 3500s. Its standard equipment included air-conditioning, sunroof and headlamp wash-wipe. The air-conditioning unit had been designed ultimately to appear on North American SD1 cars, and was probably introduced on home market cars to ensure that any problems could be sorted out before the cars were launched there.

The new car was mechanically identical to the 3500 cars, and came with five-speed manual transmission as standard, or optional automatic transmission for £230 (in practice, most V8-S cars were automatics). However, it was slightly slower because of the 340lb extra weight it carried, mainly due to the additional equipment fitted. It could also be less economical on fuel, especially if the air-conditioning was used much (the ventilation system had a setting marked 'Econ', to remind drivers that fuel consumption would be far better with the air-conditioning not in use). *Autocar* found when they tested a manual V8-S car that the 0–60mph time was over a second slower than the 3500 (9.7 seconds), and overall fuel consumption was lower, at 19.3mpg.

Metallic paint was available as a no-cost option, so most V8-S cars came with this. Early colours were blue, gold and what *Autocar* magazine described as 'a rather startlingly bright but certainly distinctive green'. Triton green was a common V8-S colour, also used on Triumph sports cars. Later on, the V8-S was available in black (for £38.96 extra) as well as other standard 3500 colours.

The Rover V8-S – the Luxurious SD1

The most obvious way of distinguishing the new luxury model was by the V8-S badge at the rear of the car.

Externally, it was easy to distinguish the V8-S from other 3500 cars. Alloy wheels were standard – on the first 900 cars, these were painted gold. Black bumpers and double coachlines identified the car, together with a small intake under the bonnet line which allowed extra cooling with the air-conditioning unit fitted. The model had a zone-tinted windscreen, as well as chromium-plated door-handles and exhaust tailpipe. The traditional Rover viking ship symbol (much loved by Rover buyers) reappeared on the bonnet badge, replacing the skeletal design. Rover V8-S badging included stylized 'Euroscript' lettering at the rear, and a V8-S badge, together with 'V8' motifs on the front wings.

Inside, there were deep, shag-pile carpets, cross-ribbed velvet-covered seats and cloth-covered door facings. Leather upholstery was available as an option for £289.60. The V8-S had front and rear head-restraints with detachable headrest cushions. The primary colour for all V8-S interiors was a rather drab shade of Nutmeg (brown), with seats and other fittings in beige or bronze. Detailed changes to the interior included brightwork surrounds on the console fascia, air vents and door-locks.

The V8-S was the first car to be fitted with SU-Butec's new air-conditioning control unit, which was positioned at the head of the centre console. The sliding steel sunroof was manually operated and could be locked at the desired position. Headlamp wash-wipe was operated by the windscreen-washer stalk when the headlamps were switched on.

Like the Rover 3500, the V8-S had electrically operated windows, central door-locking and a radio and stereo cassette player with front and rear speakers as standard.

The V8-S had twin gas struts to lift and retain the bonnet – these replaced the mechanical bonnet stays fitted to other Rovers. A larger, five-litre capacity washer bottle in the engine compartment coped with additional demands from the headlamp wash-wipe.

Generally, the better-specified SD1 received a warm welcome from the motoring press. *Autocar* believed the specification changes gave the car 'a little extra prestige appeal'. The improved interior was definitely approved of – *Autocar* liked the inviting, thick shag-pile carpet and velvet seats, which ' gave a much more sumptuous look as well as adding to comfort'. They commented that the front seats provided 'the sort of firmness in the small of the back area that helps for correct posture'. *Car* magazine also approved of these changes, while noting that:

> Safety requirements, not to say fashion, dictated the upper-Euro decor rather than traditional English as on Rovers of yore – which would no doubt have been preferred by many impressionable buyers with an eye for polished timber.

However, a number of road testers commented that the brown colour of the interior, together with all the large headrests and front seat cushions in place, gave them a claustrophobic feeling, as well as causing a loss of all-round vision, particularly to the rear.

The new heating/air-conditioning unit was found to work extremely well. *Autocar* commented that '... the terrific amount of cooling available when wanted, proved most impressive'. The cars tested by the motoring press were also found to be much better put together than earlier SD1s. *Motor* said:

> If our test car was anything to go by, Rover have now achieved the standard of finish that the excellence of the basic design deserves. The car's structure feels rattle-free, and it looks good with metallic paint and gold-painted alloy wheels.

There were very few criticisms, and these were minor ones. *Autocar* said:

> For aesthetic reasons ... there is no wiper for the big and very gently sloping rear window. Rover's designer, David Bache, says that rain blows off it as soon as speed gets over about 30mph, which is true enough but still means that in dense city traffic one can be driving for a long way in the wet with no rear vision other than through the mirrors.

V8-S standard upholstery was cross-ribbed velvet. The customer could order leather as an extra-cost option.

The Rover V8-S – the Luxurious SD1

Motor noted that:

> The V8-S is disconcertingly sensitive to crosswinds at high speed, detracting from the car's otherwise superb cruising ability. No doubt the addition of spoilers front and rear would afford an easy solution.

However, *Motor* generally liked the new V8-S model very much, remarking:

> Happily for BL, the Rover as a basic design is still an impressive all-rounder. Refined, versatile and blessed with excellent road manners, it is virtually unrivalled as an economical long-distance express … Overall, however, it is gratifying to find that the fine qualities that made the Rover Car of the Year … remain essentially undiminished four years later. Few cars are as relaxed and peaceful as the Rover is at high cruising speeds, thanks not only to the high gearing, but also the suppression of wind noise, which is much better than in earlier cars.

They concluded:

> For refinement, it compares well with its illustrious competitors, even at its somewhat daunting price of £11,852, which is nearly £1,700 more than that of the basic 3500. If you want automatic transmission, that is a mere £262 extra, but what a shame to lose that gloriously high-geared fifth speed, which is one of the greatest attractions of the car!

Autocar compared the Rover V8-S favourably with its competitors, saying that it 'remains an extremely good car, ahead of its rivals in many respects … the Rover's appeal is for the refinement and effortlessness of a big and unstressed V8.'

There was an even rarer specified version of the Rover V8-S model, the chauffeur-driven limousine. This car had a number of features which would be appreciated by cabinet ministers and captains of industry, and was designed to be driven by a chauffeur, with the owner seated in the back of the car. It had glass partition sliding windows between the front and rear seats. There were two writing tables on either side in the back of the car, with a third slide-out table in the middle, and reading lights for the benefit of occupants. It is not known how many of these specified versions of the V8-S were actually made, nor how much they cost, as the British Motor Industry Heritage Trust has no records of these cars.

Rover V8-S compared

	Price (£)	Max speed (mph)	Acceleration 0–60mph (sec)	Overall mpg
Rover V8-S	10,699	123	9.7	19.3
Ford Granada 2800i Ghia	10,018	117	8.9	20.8
Mercedes-Benz 280E (auto)	12,351	116	11.0	16.9
Opel Senator (auto)	11,365	119	9.2	18.6
Renault 30TX	8,712	115	10.3	19.5
Volvo 264 GLE (auto)	9,995	104	12.7	18.6

(Source: *Autocar*)

The Rover V8-S – the Luxurious SD1

Every SD1 should have one! The chauffeur-driven limousine was an obscure attempt to attract sales from the lower end of the carriage trade. This was possibly the only SD1 car ever to be fitted out in this manner.

8 Rover 2300, 2600 and 3500 Models – 1980–1986

Quality control was much improved on the later cars, as had been seen in road test reports of the V8-S models. From spring 1982, all Rover SD1 cars were built at Cowley.

That September, Rover announced a newly revised range of cars: the 2300, 2300S, 2600S, 3500SE and a new top of the range model, the Vanden Plas, a name which had not been used on any previous Rover model. The main differences could be seen in the interiors of the new cars. There were four new interior colours – blue, green and light and dark brown, although not all colours were available on all models. The interiors were colour-keyed, from the instrument panel to the boot. There were velvet seats and better carpets, a leather-rimmed steering wheel and a fully carpeted rear parcel shelf, as well as illuminated switches. Outside, there were new paint colours and different door mirrors with bright instead of matt black finish, and all models had a front air intake slot. There were stainless steel tread strips, which carried the Rover name, for the door sills. The expensive Ten-Twenty safety windscreen was replaced on all models by a cheaper, ordinary laminated one.

Also, changes had been made to improve ride and noise suppression. The rubber bushes uniting the ends of the anti-roll bar to the single lower links had been given more horizontal compliance, to aid in the suppression of road noise and to reduce ride harshness. Some fine-tuning of the Boge Nivomat self-levelling struts softened the ride at the rear, without having any adverse effect on the car's handling. Considerable attention was paid to reducing road noise – sound deadening material was used between the carpets and the floor, and under the bonnet. Fuel consumption was improved by making the fifth gear ratio higher on the five-speed gearbox.

The front seats had an adjustable lumbar support. A new handwheel adjustment for backrest angle replaced the former notched lever adjustment. Seat frames were redesigned, and by reshaping the rear seats and the backs of the front seats, a little extra head and leg room was obtained for those in the back.

Ventilation was improved by the provision of a ram air facility. Models with automatic transmission got a new transmission selector with a push-button release.

There were a number of improvements to the base 2300 model, including full-width arm-rests, and lights instead of reflectors in the front doors. However, it came without adjustable lumbar support in the seats, and only had a rubber mat in the boot. It also lacked the power assistance which was standard on the 2300S cars. The new 2300S model came with central locking, lumbar seat adjustment, halogen headlamps, intermittent wipe and woven velvet upholstery. The 5.5in wide wheels of the 2300 were replaced by 6in ones on the 2300S model. The four-speed gearbox was standard on both these models, but five-speed or automatic options were available at extra cost.

Rover 2300, 2600 and 3500 Models – 1980–1986

In September 1982 Rover announced their first revisions to the SD1 range. The car left has a V8 engine, whilst the car below has a six-cylinder engine – its alloy wheels were an optional extra.

Prices of Rover SD1 range – September 1980	
Rover 2300	£6,498.27
Rover 2300S	£7,498.67
Rover 2600S	£8,375.50
Rover 3500SE	£10,374.06
Rover Vanden Plas	£11,851.61

(Source: *Autocar*)

This picture of a late Series II 2300S model comes from a sales brochure – it has more basic wheel trims than the dearer models in the range.

Replacing the 2600, the new 2600S model had the improvements for the 2300S, plus a lot of extra equipment. A sliding steel sunroof with automatic wind deflector, central locking, electric window lifts all round and self-levelling rear suspension were all standard. The five-speed gearbox was standard, and had an even higher fifth gear giving 26.3mph/1,000rpm. Automatic transmission and air-conditioning were extra-cost options.

These show the interior and instrument panel of a late Series II 2300S model, which was not as spartan as you might think.

Rover 2300, 2600 and 3500 Models – 1980–1986

The V8-S model was replaced by the 3500SE model, and this was better-equipped than the V8-S. As well as the 2600S features, it had rubber bumper inserts, mudflaps, tinted glass, and a body coachline (like that on the V8-S), silver-painted alloy wheels and a radio and stereo cassette player with four speakers.

The new top-of-the-range model was the Vanden Plas. This model was mechanically identical to the 3500SE, but was more luxuriously fitted out. It came with Connolly leather upholstery (with hairline woven velvet as a no-cost option). It had an electric sliding steel sunroof, and electric remote adjustment and heating for the door-mounted mirrors. Jet washers were fitted for the headlamps, and the radio/cassette unit had stereo on both functions. Front and rear head restraints had detachable cushions. Cruise control was standard.

This 2600S badge comes from a revised SD1 – it changed on the later, Series II, cars.

A publicity photograph for the early Vanden Plas model. Vanden Plas was originally a Belgian coach-building concern, whose British factory in Northern London was bought by Austin in the early post-war years.

Rover Vanden Plas compared (in March 1981)

	Price (£)	Max speed (mph)	Acceleration 0–60 mph (sec)	Overall mpg
Rover Vanden Plas	12,475	123	9.7	20.7
Audi 200 Turbo (auto)	12,950	123	8.7	17.7
BMW 732i	14,325	127	8.0	19.8
Mercedes-Benz 280E (auto)	12,775	116	11.0	16.9
Opel Senator (auto)	12,352	119	9.2	18.6
Volvo 264 GLE	11,104	111	10.8	18.7

(Source: *Autocar*)

However, air-conditioning (previously standard on the V8-S) was now an optional extra for additional cost.

The Vanden Plas model came with automatic transmission as standard, but the five-speed manual gearbox was available as a no-cost option. On both this and the 3500SE, the higher fifth gear ratio, in conjunction with the 3.08-to-1 final drive ratio of the V8 models, gave an exceptionally high overall gearing, providing no less than 29.7mph/1,000rpm, or 100mph cruising at only 3,350rpm, (which made the V8 cars the highest-geared production cars in the world at the time). However, the Vanden Plas was 165lb heavier than the old V8-S and was therefore slower in acceleration times than the 3500SE.

Dunlop Denovo wheels and tyres were available as an optional extra on the 2300S and 2600S, or at no extra cost on the 3500SE and Vanden Plas. The bodies were now painted by the cathodic process – with the body acting as the cathode and not as the anode, as was previously the case. This change brought about a considerable improvement in the corrosion protection the body received, and was effective in preventing rust spreading from stone chippings.

Motor tests generally welcomed the improvements to the new models. *Autocar* tested a Vanden Plas model in March 1981, and commented:

> We came to like the Vanden Plas very well indeed while it was with us for test, but we are not fully convinced that the extra cost over a 3500SE is fully justified.
>
> ... It's a pity that the Rover is not rather more refined in its ride comfort, and although perfectly acceptable on good roads it does not offer the smoothness of travel enjoyed in the BMW or, to a lesser extent, in the Opel Senator and Mercedes 280E. But it has redeeming features which make it a very strong competitor in the class, and has the distinction of being the best-equipped car of them all. Add to this, the fact that it has a much more useful potential for carrying awkward loads, and that it proved the most economical of the group, and one begins to see sound reasons for choosing the Rover Vanden Plas. It's not the best all-rounder, but it's the one that comes out on top when all aspects are weighed up.

The new cars were promoted with the slogan 'One success leads to another'. Adverts stated:

There are now 5 new Rovers. That's a direct result of the success of the original award-winning Rover concept – a combination of performance, luxury and outstanding fuel economy ... All five new Rovers are exceptionally well equipped – power steering for example on the 2300S ... And, in the new Rover Vanden Plas, you'll find ultimate luxury with cruise control, automatic transmission, electrically controlled sunroof and door mirrors, and Connolly leather seating ... Now there's a Rover for every need, budget and lifestyle.

Much more radical changes were imminent, however. On 20 January 1982 Rover launched a facelifted SD1 range in the UK, which included the same five models previously mentioned. It also included a totally new model, the four-cylinder 2000, which will be looked at later on. These changes cost the company relatively little by motor industry standards (between five and seven million pounds) and gave the SD1 cars a new lease of life which lasted until 1986. At this point, it was already known that the SD1 cars would be replaced by the joint Rover-Honda project, codenamed XX. The new Rover models were marketed in Europe, following their debuts in Switzerland at the Geneva Motor Show in March.

Generally, the motoring press thought the developments to make the SD1 cars better were beneficial. In the words of *Motor*:

> Overall, though, BL have responded well to criticisms, complaints and suggestions with a worthwhile package of improvements which make a fine range of cars even better than before.

The equipment of all SD1 models now included, among many other items, an improved screen wash system, twin fog warning lamps, a radio (the specification of which varied with the model), a central armrest in the rear seat, twin remote control door mirrors, a steering wheel adjustable for both reach and rake, and rheostat-controlled instrument illumination. The most expensive Vanden Plas model was fitted with an electric sun roof, cruise control, an FM radio/stereo cassette player, and a

Rover SD1 range compared (in 1982)

Model	Engine capacity (cc)	Max power (bhp/rpm) (DIN)	torque (lb ft/rpm) (DIN)	mph/1000rpm (top gear)
2000	1,994	101/5,250	120/3,250	23.3
2000 auto	1,994	101/5,250	120/3,250	20.8
2300	2,350	120/5,250	134/4,000	26.3
2300 auto	2,350	120/5,250	134/4,000	20.8
2300S	2,350	120/5,250	134/4,000	26.3
2300S auto	2,350	120/5,250	134/4,000	20.8
2600S	2,597	132/5,000	152/3,750	26.3
2600S auto	2,597	132/5,000	152/3,750	23.3
3500SE	3,528	157/5,200	198/2,500	29.7
3500SE auto	3,528	157/5,200	198/2,500	23.5
Vanden Plas	3,528	157/5,200	198/2,500	29.7
Vanden Plas auto	3,528	157/5,200	198/2,500	23.5

Smiths trip computer. The Denovo tyre option had been deleted by this time.

The new updated cars had a number of styling changes, which made them easy to recognize – you could not mistake the new models for the old ones. They are generally known as Series II cars. The basic, well-liked shape of the car was kept, but one of the biggest changes was the altered shape of the rear of the car, which now appeared with a much larger back window – this was now 6in (15.3cm) deeper and was fitted with a rear wiper. Other changes included styling changes to the front of the car, which included flush-fitting halogen headlamps and a full-width grille over a revised radiator air duct above the front bumper. There were new moulded black plastic bumpers, which were designed to shatter on heavy impact, rather than push back and damage body panels (in practice, though, Rover owners found that these had a tendency to be damaged if you so much as looked at them)! Higher specification models had stainless steel trim on their bumpers. There was a large, colour-keyed front spoiler on all models from 2300S upwards, which improved stability at high speeds, as well as fuel consumption. New, multi-spoke alloy wheels were fitted to the two V8 models – the two S models had new stainless steel wheel trims, whilst 2000/2300 cars had polycarbonate ones.

The new paint colours were deliberately changed to be more like the conservative Rover colours of old, although Rovers pre-SD1 had never had metallic paint. On early Rover SD1s, certain metallic colours like silver and gold were dull, because of a lack of familiarity with the new paint system used. However, by 1982, knowledge of the paint system had improved dramatically, and this is why the Series II cars had more brilliant-looking metallic paint colours.

Graham Lewis was involved with the interior design of the Series II cars. He remembers that the facelift was intended to put right the earlier quality control problems, and to cure any squeaks and rattles. The door facings were redesigned – customers had demanded the return of wood and leather, but the design team had to put these ideas into practice without spending too much on any changes made – therefore, much of the interior was carried over from the Series I cars. The seats had to be changed because box-pleated seats were difficult to make, and these were restyled with more cloth content. Changes to the fascia moulding were slight, apart from the addition of wood on the more expensive models. The instruments were revised, to give the car a more modern feel. However, the speedometer and tachometer had to have quadrant-shaped – scales with the restyling of the instrument cluster, space was not available for round instruments, as found on the Series I cars.

The new instrument panel had an electronic speedometer, tachometer and low oil level warning light. A trip computer was fitted as standard to the Vanden Plas and was available as an option on the rest of the range. There was also a different steering wheel, of thicker-rimmed design. Models not fitted with a computer had a digital clock with a stop-watch facility. There were slide and turn knob controls for the heating/ventilation system, a new gear lever and gaiter on manual gearbox cars and colour-keyed seat belts. The centre console now carried an ashtray, lidded box or cassette holder, a thumb wheel for front/rear speaker balance control where applicable, and a keyboard for the stop watch or trip computer facility.

New door facings incorporated window sill mounted door lock buttons. Those in the front doors acted as master switches for the central locking system, which was a standard feature on models from the 2300S to the Vanden Plas.

Rover 2300, 2600 and 3500 Models – 1980–1986

These cars clearly show the changes on the Series II SD1 models. Most obviously, one can see the large front spoiler and the much deeper rear window. The headlamps and wheels were also changed.

Head and leg room were slightly improved for front seat occupants (although, in practice, most people failed to notice any obvious difference). Luxurious hair-line velvet upholstery, previously a no-cost option on the Vanden Plas, became standard on the 3500SE which, in common with the Vanden Plas, also boasted electrically controlled and demisted door mirrors and an electric radio aerial. The 2300S gained a steel sliding roof as standard, a feature which was already fitted to models higher in the range.

(Above) *The final assembly area at Solihull: judging by the apparent lack of activity, was this photograph taken during one of BL's infamous periods of industrial action?*

An early publicity shot which proved that some of the new Rover SD1's electrics worked at least some of the time!

(Above) *Early interior view showing the marked departure from traditional Rover values – former P4 and P5 owners could well have found this shocking.*

Because the V8 engine is relatively wide, under-bonnet access for the in-line engines fitted to the lesser models was particularly good, as this unusual overhead view shows.

(Below) *Early Rover 2300: note the lack of a mirror on the nearside of the car (one of the areas of cost-cutting to produce the then entry-level SD1).*

(Right) *Even minor items of trim, such as the 'Rover' name badge moulding of this Series I car, changed on the Series II car.*

A comparison between the recessed headlamp assemblies of the Series I car (left) and the more aerodynamically efficient, flush-mounted lights of the Series II (below), together with the more boldly coloured indicator lens. Note the additional brightwork – perceived by marketing departments as more up-market.

(Top) *Driving the six-cylinder-engined 2600 on the open road.*

(Above) *On the Series I car, the sill of the tail-gate window was higher than that of the Series II, more nearly reflecting the Pininfarina-inspired styling.*

(Left) *Of the early cars, the 3500 and 2600 had these 'door open' warning lights fitted to the front doors only. Safety pundits would say that all four doors should have been so fitted.*

A Series I door mirror: being black-coloured mouldings, these did not help to promote the up-market image that the Rover name would suggest. These were changed to the bright metal type from very late Series I cars onwards.

(Above) *This view of a Series II 2300S model shows how Rover attempted to move away from a very basic specification for the lower end of the SD1 range.*

The prominent air dam on most Series II cars is clearly shown in this picture of a 2300S. On Vanden Plas models an ancillary light was fitted outboard of each jacking point under the front bumper – the jacking points being concealed by the under rider mouldings.

(Top) *As can be seen from this picture, with tow-bar attached the SD1, like its predecessors, was a popular choice with caravan owners, as the good power-to weight ratio (particularly of the V8-engined models) made it ideal for long-distance towing.*

(Above) *This late Vitesse is displaying the fairly unusual colour choice (for an SD1) of black.*

(Left) *The cross-spoked alloy wheels of the Vitesse, whilst looking attractive, were tedious to keep as clean as this.*

(Above) *This immaculate Vanden Plas EFi is an example of one of the rarer Rover SD1 variants. This car, in Ferrari-like bright red, accentuates David Bache's styling inspired by the Pininfarina-styled Ferrari Daytona.*

The fuel-injected engine of the Vanden Plas EFi is the same unit with automatic transmission as was fitted to some of the earlier versions of the Vitesse.

(Right) *This is the leather interior of the Vanden Plas EFi model. The booster cushions fitted to the head restraints are the most obvious difference from the standard Vanden Plas interior, other than the leather seat facings – these were standard on the Vanden Plas EFi, but were extra cost items on Series II Vanden Plas cars.*

(Above) *The two-tone colouring of the revised door interior pads attempted, in conjunction with the bright trim lines and wood-finished fillets, to give the interior a more luxurious finish than had been the case with the Series I cars. In this view of a driver's door, the revised position of the electric window control switches is shown. On Series I cars, they had been on the central console behind the gear stick – a position found to be vulnerable to malfunctioning due to dropped dirt such as food crumbs and cigarette ash.*

(Below) *Some of the various models of the SD1 that have been available comprising plastic construction kits and die-cast miniatures. The red model in the right foreground is a comparatively expensive, hand-finished pewter model of the actual Vanden Plas EFi illustrated elsewhere.*

Of more importance to traditional Rover buyers, the cheap and stark interiors of earlier cars was all but forgotten – walnut trim panels were fitted on the more expensive models (2600S and upwards). The S models had satin-finish wood door inserts, while the 3500SE had burr walnut. The Vanden Plas model had the most wood – it had burr walnut door inserts and a burr walnut wooden panel across the fascia.

The Vanden Plas model came with leather seats as standard (with velvet material a no-cost option). It also had pockets in the backs of the front seats, rear seat head restraints and detachable cushions (on the head restraints). 'Clearcoat' metallic paint was a no-cost option for the Vanden Plas model, but was also available on the lesser models for an additional charge.

Rover claimed that the five-speed manual 2600S model had a 116mph maximum speed, and a 0–60mph time of 10.3 seconds, with the manual five-speed 3500SE model reaching 126mph top speed, and attaining 0–60mph in 8.6 seconds.

The new range of SD1 cars cost from £7,450 for the cheapest model, the 2000, to £14,787 for an automatic Vanden Plas (Ford Granada prices were not much different at the time, starting at £7,211).

Many of the changes had been carried out in direct response to customer demand. The poor rearward visibility had been commented on in numerous road tests of earlier SD1s. One motoring writer had his prayers answered when he wrote in 1980:

> There is one other major styling change which I would ask for, but am unlikely to get in view of the cost if nothing else – and that is to straighten out that silly upturn of the window waist line towards the rear, which ends up in the dangerously high bottom edge to the back glass – hiding too much to the rear, particularly when you're reversing in a street perhaps running with unguarded children.

Styling changes for the new cars made use of the opportunity to improve this problem. (It is believed that there may have been a limit

This picture shows the interior and instrument panel on a 1984 Vanden Plas. Leather seats (as shown here) were available at extra cost. The return of wood and leather was much appreciated by traditional Rover owners.

on glass size when the car was first made in 1976). The larger rear window improved rearward visibility, and therefore improved safety and made parking less hazardous.

An automatic choke was adopted on all models, and a five-speed manual gearbox was standard on all manual cars. A three-speed Borg Warner Model 66 automatic transmission was fitted as standard to the Vanden Plas, and was optional on all other models. Mechanically, the new models differed little from the old ones. There were some improvements, including modifications to the brake system to improve pedal feel and stopping power – changes included a larger servo and master cylinder. Models with self-levelling rear suspension (the 2600S, 3500S and Vanden Plas) had revised settings for the Nivomat self-levelling units to improve ride and handling characteristics. All models had a higher output alternator. The V8 models now had twin Solex 175 CDEF carburettors, which improved fuel economy, and the V8 engine had a stiffened cylinder block which reduced noise. The six-cylinder engines now had a remote air cleaner to reduce induction roar. The final drive ratio on the cars ranged from 3.9 to 1 for the manual 2000, to 3.45 to 1 for the six-cylinder cars, and 3.08 to 1 for the V8 models.

A more dubious change was the extension of servicing intervals, to reduce ownership costs. It has to be said that, in practice, this was a response to the current fashion of the time, and for no other reason. Other manufacturers were using the new 12,000 mile service intervals on their cars. On the new Rover cars, the servicing intervals were increased from 6,000 to 12,000 miles, with the deletion of the 6,000 mile oil change. Contract breaker points and sparking plugs were to be cleaned at 12,000 miles and replaced at 24,000 miles. Brake wear sensors were fitted to all models. Rover literature boasted how the new Rovers reigned supreme over their competitors in terms of running cost:

Rover servicing costs (mandatory parts plus labour) over the industry recognized 50,000 mile/four year period range from £236 for the new 2000 to £308 for the V8

Official fuel consumption figures (mpg)

Model	Urban	Steady 56mph	Steady 75mph	Price (£)
2000	23.9	42.6	32.7	7,450
2000 auto	24.7	36.2	27.6	7,757
2300	20.5	40.1	30.8	7,970
2300 auto	21.0	34.6	24.6	8,277
2300S	20.5	40.1	30.8	9,359
2300S auto	21.0	34.6	24.6	9,666
2600S	19.0	40.9	31.1	10,177
2600S auto	20.2	37.4	28.0	10,484
3500SE	16.9	38.8	28.0	12,546
3500SE auto	18.5	31.9	25.6	12,852
Vanden Plas	16.9	38.8	28.0	14,480
Vanden Plas auto	18.5	31.9	23.6	14,787

(Source: *Autocar*)

models. This compares, for example, with £283 for the 2 litre Renault 20 TS up to a massive £511 for the Volvo 244 DL.

... On fuel economy too, the Rover range is way ahead of its competitors on most statutory consumption cycles.

BL Cars stated that one in two Rover buyers chose to buy the same marque again (one wonders what the other buyer chose instead?) According to company claims, out of 22,000 Rover cars sold in Britain in 1981, 80 per cent were sold to business users, with over 70 per cent of professional men, directors and senior managers choosing the 3500/2600 models as their first preference. The company claimed that Rover's share of the executive car market increased from 12.6 per cent in 1980 to 15.1 per cent in 1981.

Harold Musgrove, chairman of BL Cars at the time of the launch of the Series II SD1 cars, was quoted as saying:

> We've gone for improved refinement, up to the minute styling features, low running costs and even better quality on the latest Rovers (it could not have been much worse than it had been on early SD1s).
>
> ... We aim to preserve marque loyalty as well as winning new customers with the new Rovers. In our book, no other executive car can match them in term of specification and running costs.

Generally, when it came to advertising the new restyled models, the company promoted the whole range of SD1 cars. However, advertisements made the most of promoting the return of traditional Rover virtues on some models. In 1982, advertisements exhorted prospective buyers to:

> Experience the sensation of power, the luxury of walnut and the prestige of Rover ... as you sit back in supreme comfort, you'll appreciate that a traditional Rover luxury – walnut panelling – has made a welcome return ... Ask your dealer for a test drive. And enjoy the advanced driving experience.

When *Motor* carried out a long-term test of a 3500SE model, they found that it gave little trouble and seemed basically sound (the only obvious delivery fault was a loose surround to the locking knob on the driver's door). They noted the improved quality, which seemed to be reviving the popularity of the SD1 cars.

As far as British Leyland/Jaguar Rover Triumph was concerned, the company was called Austin Rover from the middle of 1982.

There were few major changes in the specifications of these cars between the time of their introduction and the end of production in 1986, although there were several interior changes and a number of improvements. In 1984, the door handle surrounds changed from bright metal to black finished ones. From May 1984, the rear badges were mounted on the black plinth, and door mirror bases were reshaped to cut down on wind noise – also, car entertainment systems were upgraded across the range. From July 1985, external radio aerials were replaced by an aerial which was wired into the rear window heating element. In January 1986, small square repeater flashers were added to the front wings of all models.

From the middle of 1983, the original solenoid-operated type of central locking system was replaced by a more reliable motorized one. Improved air-conditioning and cruise control systems were introduced in 1984, and the ICE systems of all SD1 cars were upgraded in May 1984. Bronze-tinted glass was standardized across the range of SD1 cars from October 1982, and electric mirrors from 1985. In 1984, both V8 and six-cylinder cars were given SU HIF 44E carburettors. From 1983, automatic models had GM 180 transmissions, and gear ratios were slightly altered.

There were several changes in the type of cloth seating material used on SD1 cars from 1983, mainly to keep this in line with the colours and fabrics used on other Austin Rover models. 1984 season Vanden Plas models had Rascelle knitted fabric upholstery as an option instead of leather, but from 1985, box velvet material became standard on Vanden Plas cars (with leather an extra-cost option). For 1983 season Vanden Plas and SE models, a carpeted kick-panel was added to the door trims, and from 1984 all door trims had a panel in a contrasting colour.

In October 1982, a 2600SE model was introduced with specification similar to the 3500SE. In May 1984, both the 2600SE and 3500SE models were discontinued, and a new 2600 Vanden Plas model was introduced, with similar specification to the 3500 Vanden Plas. Production of the other SD1 models continued until early 1986.

The rear badge on SD1s was the main way that non-SD1 enthusiasts were able to differentiate between the various models.

9 The Rover 2000 – a Famous Name Revived

The launch of the facelifted Rover SD1 range in January 1982 saw the addition of a new model, and the revival of a famous name – the Rover 2000.

Austin Rover publicity material stated that:

> The market for cars of between 1850cc and 2150cc capacity is the fastest-growing in the executive sector, having risen by an estimated 10 per cent in the last 5 years. The Rover 2000, therefore, with its attractive blend of value, accommodation and style, is in an ideal position to challenge for an increasing market share.

They also revealed why a 2000 model had returned:

> The ... Rover 2000 ... takes the SD1 range directly into the fast-growing 1850cc to 2150cc executive car sector currently dominated by the 2-litre Ford Granada.

Ford were selling a lot of 2-litre Granadas to the important fleet sector. In 1981, 25,214 Ford Granadas were sold, compared with 21,504 Rovers. At this level, the Rover also competed with the 2-litre Ford Cortina. The 2-litre executive class was dominated by Ford, Mercedes, Citroen and Renault.

The Rover 2000 was very different from its earlier namesake, being an SD1 fitted with the two-litre version of the Austin/Morris 'O' series engine.

Rover 2000 (1981–85) – specification

Engine Front; rear drive
Head/block Al alloy/cast iron
Cylinders 4, bored block
Main bearings 5
Cooling Water
Fan Electric
Bore, mm (in) 84.4 (3.32)
Stroke, mm (in) 89.0 (3.50)
Capacity, cc (cubic in) 1,994 (121.68)
Valve gear Ohc
Camshaft drive Toothed belt
Compression ratio 9.0 to 1
Ignition Contact breaker
Carburettor Two SU HIF 44
Max power 101bhp (DIN) at 5,250rpm
Max torque 120lb ft at 3,250rpm

Transmission
Type 5-speed manual
Clutch diameter 8.5in
Actuation Hydraulic

Gear	Ratio	mph/1000rpm
Top	0.79	23.35
4th	1.00	18.45
3rd	1.39	13.27
2nd	2.09	8.83
1st	3.32	5.56

Final drive gear Hypoid bevel
Ratio 3.90-to-1

Suspension
Front – location MacPherson struts
    – springs Coil
    – dampers Telescopic
    – anti-roll bar Yes
Rear – location Live axle on trailing arms, torque tube plus Watts linkage
    – springs Coil
    – dampers Telescopic
    – anti-roll bar No

Steering
Type Rack and pinion
Power assistance Optional
Wheel diameter 15.5 x 16.5in
Turns lock to lock 3.8

Brakes
Circuits Dual, split front/rear
Front 10.5in diameter disc
Rear 9.0in diameter drum
Servo Vacuum
Handbrake Centre lever acting on rear drums

> **Rover 2000 (1981–85) – specification** *(continued)*
>
> **Wheels**
> Type   Pressed steel
> Rim width   5.5in
> Tyres – make   Michelin XVS
>   – type   Radial ply
>   – size   175HR-14
>   – pressures   F 28, R 30psi (normal driving)
>
> **Equipment**
> Battery   12V35Ah
> Alternator   65A
> Headlamps   110/230W
> Reversing lamp   Standard
> Electric fuses   18
> Screen wipers   Two-speed plus flick wipe
> Screen washer   Electric
> Interior heater   Air blending
> Interior trim   Cloth seats, cloth headlining
> Floor covering   Carpet
> Jack   Screw pillar
> Jacking points   One at each corner
> Windscreen   Laminated
> Underbody protection Cathodic paint treatment, wax-injected box members, pvc underseal
>
> (Source: *Autocar*)

Not everyone at Rover, however, was endeared with the idea of a 4-cylinder addition to the SD1 range. Spen King can remember not wanting a 4-cylinder model, because he felt this was down-market (the motor industry had moved on somewhat since the original Rover 2000 model was launched in 1963). The SD1 version of the Rover 2000 was less powerful than the sporting 2000TC model of the P6 range had been (*see* figures at the end of this chapter).

When looking for a 2-litre engine, the company decided not to make a smaller version of the 2300 six-cylinder engine, as they wanted a lighter, simpler unit. They looked around the company's existing engine stock, and chose to uprate the 'O' series engine that was fitted to the Morris Ital and Austin Princess models. When installed in the new Rover, the engine looked somewhat lost – but at least there was good access to work on it.

(*Autocar* joked that the four cylinder Rover had extra luggage space under the bonnet!)

The 'O' series unit, in Rover 2000 form, had twin SU sidedraught carburettors. A new bell-housing was needed to mate with the five-speed, 77mm gearbox. A new fabricated exhaust manifold and a remote air cleaner were required, as well as a different radiator with an electric fan mounted ahead of it. Surprisingly, a larger diameter propshaft than that used for the six-cylinder and V8 models was specified, in order to cut down resonance.

The 2000 came with the five-speed gearbox as standard – Borg Warner Type 66 three-speed automatic transmission was an optional extra (in practice, not many automatic 2000s were sold). In order to overcome the poor power-to-weight ratio of this unit in the heavy SD1 body shell, a lower final drive ratio of 3.9 to 1 was fitted, which gave high final gearing of 23.3mph per

The Rover 2000 – a Famous Name Revived

The four-cylinder 'O' series engine installed in the SD1 body had almost an excess of empty space around it with the bonus of easy access for maintenance.

1,000rpm in top with the five-speed, manual gearbox. The 1,994cc engine developed 101bhp at 5,250rpm, and 120lb ft of torque at no more than 3,250rpm. The manufacturers claimed a 0–60mph time of 12.5 seconds and a maximum speed of 104mph.

The 2000 lacked power-assisted steering and sunroof (although both were available as optional extras). As the four-cylinder engine was much lighter than the larger Rover engine, softer spring and damper rates were used at the front. There were variable rate coil springs at the rear, not the self-levelling rear damper units on some of the dearer models. The 2000 did have an automatic choke, and came with velvet upholstery, cut-pile carpet, digital clock, push-button twin-speaker radio, halogen headlamps and the rear window wash-wipe. Other optional equipment included alloy wheels, metallic or black paint, tinted glass, computer and central door-locking. Air-conditioning, however, was not available as an option. Other cost-cutting measures meant that the car only had rubber trim in the boot, plastic wheel trims, and had no spoiler or bright trim on the bumpers.

How did the motoring press rate the new Rover 2000? In the words of *Motor*, 'the Rover 2000 is reborn in SD1 form: bad news for the 2-litre Granada?' The Rover 2000 was the first facelifted Series II model to be thoroughly tested by the motoring press. Motor magazines, by now thoroughly accustomed to the far more powerful six-cylinder and V8-engined versions of the SD1, were less enthusiastic about the 2000 model when they did test it, and reaction was

All Series II cars, including the base model 2000, were fitted with the rear window wash-wipe.

mixed. *Motor* magazine managed to obtain a top speed of 102mph, with 0–60mph being reached in 12.4 seconds; they remarked that both performance and fuel economy (23.3mpg overall) were poor, and that the car required a lot of gear-changing. *Motor* did, on the other hand, like the improved interior, (even if their test car appears to have been built on a Friday afternoon):

> What is undeniable is that the Rover's interior, even in base 2000 form, exudes an air of luxury. The velour-trimmed seats and thick carpets are impressive, as is the fact that the instrumentation is exactly the same as that of the top-of-the-range Vanden Plas. The overall effect is both plush and tasteful. We must be more critical, however, of the interior build quality. While the test car was with us there was a constant, and annoying, creaking from the area of the facia. Worse still, both a plastic door lock surround and seat recline handwheel fell off.

Several road tests commented on the noisy power unit – above about 3,500rpm the engine sounded unpleasantly harsh and strained, although road roar was well suppressed.

Motor concluded:

> A smarter and more effective Rover range is something of which we wholeheartedly approve. Europe's biggest and best hatchback is a commodity BL, understandably, should want to preserve. However, we find the 2000 a difficult car to come to terms with. Undoubtedly an attractive alternative to the four-cylinder Granada and, indeed, a likeable car in its own right, the crudeness of its engine seems at odds with standards of refinement and luxury attained by the rest of its design. With neither performance nor economy to commend it, we can think of no good reason for not spending another £500 for the six-cylinder 2300.

Autocar, on the other hand, was more enamoured with the car:

> Previous complaints regarding the SD1's lack of rear visibility have to a large extent been answered. The deeper rear hatch window makes the car far easier to park (particularly for the shorter driver), and once on the road the view rearwards is now much less obstructed.
> ... The Rover scores well ergonomically and on space and ventilation ... Back comes BL with a new product needing less frequent service attention, and with today's motoring backdrop, that may be the most powerful argument of all for the car, for both fleet and private buyers alike.

When compared with its rivals, *What Car?* preferred the Rover 2000 to the 2-litre Ford Granada, commenting:

> Against the Granada the Rover scores on more points than just style and practicality. It is more economical and almost as fast: it is as much fun to drive and it is better value. The Granada does, however, have more space and a smoother ride, though it is nowhere near as interesting.

When the new model was promoted, advertisements asked motorist to:

> Experience the new 104mph Rover 2000 ... The Rover name stands for innovative design, performance, prestige, technological development and achievement. Qualities that have long made Rover a very special driving experience ... The Rover 2000 comes to you with advances in styling, in comfort, in economy and in finish ... by using the world's most advanced and

Rover 2000 compared

	Price (£)	Max speed (mph)	Acceleration 0–60 mph (sec)	Overall mpg
Rover 2000 (SD1)	7,450	103	12.4	22.8
Rover 2000TC (P6)	N/A	108	11.9	22.3
Ford Granada 2000GL	8,828	102	11.9	21.2
Audi 100CL	7,653	108	10.7	21.4
Citroen CX Athena	7,634	109	12.5	26.2
Renault 20TS (auto)	7,602	100	14.8	22.9
Talbot Tagora GL	7,739	106	11.3	23.9

proven paint technology we're ensuring that the 2000 is protected against all the rigours of the British climate ... The new Rover 2000. The car that promises to be as appealing and legendary as the original.

One idea which never reached the production stage was a turbocharged version of the Rover 2000. Dennis Barbet, who was involved with engine development, can remember developing and experimenting with the turbocharged Rover 2000. The work on this was carried out by Broadspeed, who had previously carried out work on a turbocharged version of the Triumph TR7. The turbocharged Rover 2000 was the brain-child of chief engineer Graham Atkin (who had previously worked for Lotus). Graham Atkin did not succeed in selling the idea to upper management, other projects being considered more important. The turbo-charged SD1 was never produced because of the cost of development and production, but also because it would have overlapped with more powerful versions of the SD1. The 'O' series-engined MG Montego Turbo which was produced, was not particularly liked by the motoring press, being a light front-wheel drive car which appeared to have a will of its own.

There were a few changes to the specification of the Rover 2000 model between 1982 and 1986, in line with those on other Series II cars (as described previously). Austin Rover were soon under pressure to improve the model's very basic specification, and did this from October 1982, when tinted glass and central locking became standard features. As with other models, there were several changes in seat material (Shetland tweed upholstery was standard between July 1983 and May 1984, but there was a return to plain velvet afterwards). In May 1984, further improvements included the addition of stainless steel wheel trims, and bright trim on the bumpers. The 2000 model now had electric windows, walnut veneer inserts in the doors, front seats with adjustable lumbar supports, passenger grab handles and a better four-speaker stereo radio/cassette player. All these additions brought the 2000 in line with other models in the SD1 range. From October of the same year, the 2000 specification included a manually-operated sunroof, power-assisted steering, intermittent wash-wipe, and electrically adjusted door mirrors (with demisting facility). At the same time, the carburettors gained an electronic mixture control system, and electronic ignition.

The last Rover 2000 was built in 1986, making way for 2-litre versions of the Rover 800 model.

10 The 2400SD Model – the Rare Turbodiesel SD1

In April 1982, a new Rover SD1 model was launched at the Turin Motor Show. This was the first-ever Rover car to be marketed with a diesel engine, the 2400SD Turbo model.

The new Rover was aimed at the European market, where motor and fuel taxation policies in certain European countries made large diesel cars much more attractive than in Britain. For example, in Italy at the time of the car's launch, diesel fuel was only half the price of petrol, while in Holland, France and Belgium, diesel fuel was 30 per cent cheaper than petrol.

Not surprisingly, then, Italy was the main market for the new Rover – out of 351 Rover SD1 cars sold there in May 1982, 173 (49 per cent) were the new diesel model. This is the reason why the car was launched there (it was also the home of the engine supplier, Stabilimenti Meccanici VM). The rest of the cars were to be sold throughout the rest of Europe. Austin Rover did not expect to sell many of these cars in the UK. However, here the fleet buyer, who tended to run up heavy mileages, would appreciate the greater distances a diesel Rover would be able to travel on a gallon of fuel.

Rover had to look around Europe for the supplier of a diesel engine, as a suitable in-house unit was not available. It was decided that the relatively low volumes envisaged made it sensible to buy in a proprietary engine. A diesel version of the V8 engine was planned for the future (a joint project between Rover and Perkins), but engineers were never able to make this engine work properly, and the project was abandoned at the end of 1983. The company therefore turned to Stabilimenti Meccanici VM, whose factory was at Cento, twenty-five miles from Bologna, to supply a suitable 2,393cc unit. They had been making diesel engines for Alfa Romeo and other car manufacturers, as well as industrial and marine diesel engines. The 2,393cc unit was chosen for several reasons – it offered good performance, was the right size and price, and was quickly available. Austin Rover claimed that in 2-litre format, the VM engine they selected was already on the road powering the Alfa-Romeo Alfetta Diesel available in the Italian market. Rover engineers designed a new sump, oil pick-up, flywheel and clutch housing – the only major parts which were required to slot the engine into the Rover body shell and mate it with the existing transmission components.

The four-cylinder engine used in the new SD1 car was the HR492 HT. This was an overhead-valve unit, which had a Bosch injection system, and a KKK exhaust-driven turbocharger. The design had four separate indirect-injection light-alloy cylinder heads. The unit was canted slightly backwards in the car's engine bay, and had the addition of hydraulic front mountings in order to damp out vibration. A non-waxing fuel system was fitted. The engine was equipped with Lucas glow plugs to speed up cold starts, from around 25 to 30 seconds delay down to around 7 seconds. The car also had a V8

Rover 2400SD Turbo (1981–85) – specification

Engine
Head/block	Front; rear-wheel drive

Head/block: Al. alloy/cast iron
Cylinders: 4, wet liners
Main bearings: 5
Cooling: Water
Fan: Electric
Bore, mm (in): 92 (3.63)
Stroke, mm (in): 90 (3.54)
Capacity, cc (cubic in): 2,393 (146.1)
Valve gear: Ohv
Camshaft drive: Chain
Compression ratio: 20.5:1
Ignition: Bosch
Turbocharger: KKK
Max power: 90bhp (DIN) at 4,200rpm
Max torque: 142lb ft at 2,350rpm

Transmission
Type: 5-speed all synchromesh
Clutch: Single dry plate

Gear	Ratio	mph/1,000rpm
Top	0.77	23.9
4th	1.00	18.45
3rd	1.39	13.27
2nd	2.09	8.83
1st	3.32	5.56

Final drive gear: Hypoid bevel
Ratio: 3.90:1

Suspension
Front – location: MacPherson struts
    – springs: Coil
    – dampers: Telescopic
    – anti-roll bar: Yes
Rear – location: Live axle, trailing arms, torque tube, plus Watts linkage
    – springs: Coil
    – dampers: Telescopic
    – anti-roll bar: No

Steering
Type: Rack and pinion
Power assistance: Yes
Wheel diameter: 15.5 × 16.5in
Turns lock to lock: 2.8

Brakes
Circuits: Dual, split front/rear
Front: 10.5in diameter disc
Rear: 9.0in diameter drum
Servo: Vacuum
Handbrake: Centre lever acting on rear drums

Rover 2400SD Turbo (1981–85) – specification *(continued)*

Wheels
Type	Pressed steel
Rim width	5.5in
Tyre – make	Michelin XVS
– type	Radial ply, tubeless
– size	175HR-14
– pressures	F 28, R 30psi (normal driving)

Equipment
Batteries	2 × 12V, 55Ah
Alternator	65A
Headlamps	110/230W
Reversing lamp	Standard
Electric fuses	22
Screen wipers	Two-speed plus intermittent
Screen washer	Electric
Interior heater	Air blending
Interior trim	Cloth seats, cloth headlining
Floor covering	Carpet
Jack	Screw pillar
Jacking points	One on each corner
Windscreen	Laminated
Underbody protection	Cathodic paint treatment, wax-injected box members and pvc underseal

(Source: *Autocar*)

9.5in clutch and the larger-diameter 2000 propeller shaft, (to keep down four-cylinder vibrations).

Unlike other SD1 cars, the diesel had an engine oil cooler mounted ahead of the radiator, which had two electric cooling fans. It also had an enlarged water pump bypass to the heater to compensate for the lower running temperature of a diesel. Gearbox ratios were revised. The 3.9-to-1 final drive (from the manual Rover 2000) and raised fifth ratio gave a road speed of 23.9mph/1,000rpm. Suspension changes were limited to stiffened front coil springs to compensate for the diesel engine's extra weight. A second battery was added to aid starting. In all, the car weighed about 330lb (150kg) more than the Rover 2000 model.

The four-cylinder ohv diesel engine. This picture shows a complete engine, partly sectioned for exhibition purposes.

The 2400SD Model – the Rare Turbodiesel SD1

	Rover SD Turbo compared			
	Price (£)	*Max speed (mph)*	*Acceleration 0–60 mph (sec)*	*Overall mpg*
Rover SD Turbo	10,500	104	14.3	29.6
Audi 80 Turbo Diesel	7,616	96	12.8	37.3
Citroen CX25D	9,116	97	17.0	31.5
Mercedes-Benz 300D	11,400	90	20.8	27.2
Peugeot 604SR Diesel Turbo	11,085	94	17.0	27.7
Vauxhall Cavalier LD	5,807	87	17.8	38.2

(Source: *Autocar*)

Service intervals were more frequent than for other 1982 model SD1 cars. Major service intervals were at 12,000 miles, but the oil had to be changed every 3,000 miles, while the oil filter needed changing every 6,000 miles. As *What Car* noted, when they tested the car:

> The diesel formula won't suit everyone. Only the driver covering high annual mileage will be able satisfactorily to play off the saving of fuel against increased servicing.

Rover claimed that the new 2400SD model was the fastest diesel saloon available in the UK at the time of the car's launch in Britain in July 1982, with a top speed of 100mph, and similar acceleration to the Rover 2000 model. The 2400SD produced a maximum 90bhp at 4,200rpm, and had good torque of 142lb ft at 2,350rpm, which gave it acceptable performance at low speeds. In addition, it returned over 30mpg.

The turbodiesel had the same high equipment levels as the 2600S model. At £10,500, it was priced between the 2600S and 2600SE models. Interestingly, although there were '2400' and 'SD Turbo' badges on the car, no badges on the car carried the

These are the badges on the 2400SD Turbo – note there was no mention of the word 'diesel' anywhere on the car.

word 'diesel'. The 2400SD came with power steering, steel sliding sunroof, central locking, electric windows, a radio, velvet seats and a colour-keyed front spoiler. Optional extras included alloy wheels, front fog lamps, electric sun roof, leather upholstery and black or metallic paint.

There were a few changes to this model after its launch. General specification changes were in line with those of other Series II cars (as described previously). During the time of production, it shared the same improvements as the 2600S model (though obviously it did not share the 2600S model's electronic mixture control system improvement during the revisions made in October 1984). The 2400SD lacked certain options such as air-conditioning, automatic transmission and the trip computer.

In their sales brochures, under the slogan 'Rover SD Turbo – Advancing The Diesel Experience', the company claimed that:

> The new Rover SD Turbo quite literally sets new standards for diesel motoring. Powered by a new generation turbocharged engine, it offers a unique blend of technology and tradition. Technology that delivers impressive performance, exceptional diesel economy and all-weather reliability. Exhaustive cold weather testing proved that the SD Turbo engine will start readily at temperatures as low as minus 20° C ... Tuned power unit mountings incorporating hydraulic damping and acoustic insulation ensure the refinements expected of a Rover.

On 6 December 1983, the Rover SD Turbo set fourteen British speed records for diesel cars, including the 5km record and the 500 mile record. The car covered a distance of over 920 miles, and was driven around the Snetterton circuit by a five-driver team. The attempt was organized by Williams Grand Prix Manager, Peter Collins. Austin Rover made use of this achievement in their advertisements.

Road tests by the motoring press (somewhat surprisingly) generally approved of the new Rover model. *Car* magazine found that, although there was a lot of noise after starting from cold:

> the diesel smooths out beautifully at medium and high revs and at cruising speed on motorways most noise is generated by the wind and tyres. The engine is only just audible.
>
> ... After a 1000km test run involving both autostrada and city driving this Rover is impressive on many fronts. It is fast, faster than the officially claimed 102mph and because of the engine's impressive torque output it has good acceleration ... From a standing start the Rover takes a few seconds to get moving but the engine revs surprisingly easily and pulls this weighty car smoothly up to a top speed which I measured at 104mph.

Motor also liked the new Rover model:

> Cruising at speed is really its forte; up to about 90mph in top the diesel quality of the engine is much less pronounced.
>
> ... Overall, though, the Rover is respectably refined, as befits its executive car image ... engine noise ... is never excessively intrusive. At motorway speeds in fifth gear it is sufficiently well suppressed to allow pleasantly relaxed cruising.
>
> Inside, the Rover diesel also fulfils its top-car image with a tasteful combination of colours and materials, including very attractive pin-striped velour trim for the seats and walnut veneer inserts on the door trims. This luxurious and well-assembled feel is followed through to the exterior which features acceptably snug panel fits and a truly lustrous paint finish.

The 2400SD Model – the Rare Turbodiesel SD1

They concluded:

Austin Rover's first diesel car for nearly twenty years lives up to its makers' 'fastest diesel' claims and also delivers the goods as far as fuel economy is concerned. The Rover 2400SD Turbo retains all the better attributes of the capable Rover range and while the driver is unlikely to forget he is sitting behind a diesel engine he can take comfort from the car's parsimonious thirst for fuel and the plushness of his surroundings. The diesel Rover will undoubtedly appeal to fleet operators looking for more economical cars for their executives; and the private owner who wants an economy luxury car, and isn't too bothered about performance, should also find the Rover a very attractive proposition.

Autocar commented that:

The Rover is the fleetest diesel car we have tested so far ... Thanks to its hatchback, it has a big advantage over the opposition on load carrying versatility, and even for European customers that should count for a lot.

The cold start control was a feature that was unique to the diesel model, as all other SD1s had automatic cold start controls, and were fitted with a coin pocket in this place in the dash.

Ultimately, however, the turbodiesel SD1 was a very rare beast, especially in the UK. Just over 10,000 of these cars were produced in total, most of which were left-hand drive models. The model was last advertised in UK catalogues late in 1985, and disappeared from car dealers showrooms soon afterwards.

An under-bonnet view of the SD1 diesel engine. This unit is unusual for a comparatively small engine, in that it has individual cylinder heads and rocker boxes for each cylinder.

11 Rover Vitesse – the Ultimate SD1?

October 1982 marked an important point in the history of the Rover SD1. At the British Motor Show that month, a brand new, high-performance version of the car was launched, called the Rover Vitesse. This new model was developed directly as the result of a successful Rover SD1 racing programme, which extended Rover's sporting image (the racing programme will be discussed in a later chapter).

Rover had originally intended to call this new SD1 flagship the Rover Rapide. However, Aston Martin Lagonda held the rights to this car name, and refused to let Rover use it on their V8-engined, high-performance large saloon. Instead, Rover named the new car the Rover Vitesse, taking the Vitesse name from Triumph cars of the past. It was an apt name, though – French for 'speed' – and the new model certainly lived up to its name.

A high-performance version of the SD1 car had been perceived by a number of people at the company as being a good idea. Spen King, for example, had pushed for some time for a high-performance version of the SD1, which would create an image that the company would be able to capitalize on. The very first Rover Vitesse was the idea of marketing man David Clarke, who had a V8-engined car built which could produce 200bhp. This particular car was effectively a racing car – it had twin-choke Weber carburettors, lowered suspension and Minilite wheels, and was green in colour. The motor sport department carried out the modifications to it. Eventually, the car was shown to the management. Harold Musgrove (who had been managing director, and was chairman of Austin Rover's Cars Division at the time of the launch of the Vitesse) borrowed the car, and became totally hooked on it. He said he wanted a 200bhp production version of the Rover SD1, but a vehicle which was much more refined and practical in everyday use than this particular one.

Further development of the Australian SD1 car's fuel injection system gave the company the opportunity it needed to improve the performance of this car. When John Davenport, the head of the company's Competitions Department, approached management and requested the production of a special high-performance, fuel-injected version of the car which would be homologated for racing for the 1983 season, his wish was granted. The idea for a high-performance version of the car therefore came partly from the motor sport department and partly from the marketing and product planning departments, who believed that this model would increase sales of all SD1 cars (they were correct in thinking this). In addition, a more expensive version of the car would improve the company's profit margins.

John Davenport recalls that a great deal of time and effort went into engineering the new model. Being in charge of motor sport, he was heavily involved, and had a major say in many aspects of the car, including the design of the plenum chamber. He says that the

Rover Vitesse (1982–86) – specification

Engine Front; rear-wheel drive
Head/block Al. alloy/Al. alloy
Cylinders 8 in 90 deg. V dry liners
Main bearings 5
Cooling Water
Fan Viscous
Bore, mm (in) 88.9 (3.50)
Stroke, mm (in) 71.1 (2.80)
Capacity, cc 3,528 (215.0)
(cubic in)
Valve gear Ohv
Camshaft drive Chain
Compression ratio 9.75-to-1
Ignition Breakerless
Injection Lucas L electronic system
Max power 190bhp (DIN) at 5,280rpm
Max torque 220lb ft at 4,000rpm

Transmission
Type 5-speed synchromesh
Clutch Single dry plate

Gear	Ratio	mph/1,000rpm
Top	0.792	29.4
4th	1.000	23.3
3rd	1.396	16.7
2nd	2.087	11.1
1st	3.320	7.0

Final drive gear Hypoid
Ratio 3.08

Suspension
Front – location Independent MacPherson strut
    – springs Coil
    – dampers Telescopic
    – anti-roll bar Yes
Rear – location Live axle, torque tube, trailing arms, plus Watts linkage
    – springs Coil
    – dampers Telescopic
    – anti-roll bar No

Steering
Type Rack and pinion
Power assistance Yes
Wheel diameter 16.5 × 15.5in
Turns lock to lock 3.0

Brakes
Front 10.1in diameter ventilated disc
Rear 9.0in diameter drum
Servo Vacuum
Handbrake Centre lever working on rear drums

Rover Vitesse (1982–86) – specification *(continued)*

Wheels
Type	Cast alloy
Rim width	6.5in J
Tyre – make	Goodyear NCT
– type	Radial ply, tubeless
– size	205/60VR-15
– pressures	F 24, R 26psi (normal driving)

Equipment
Batteries	12V 66Ah
Alternator	75A
Headlamps	110/230W
Reversing lamp	Standard
Electric fuses	10
Screen wipers	Two-speed plus intermittent/flick wipe
Screen washer	Electric
Interior heater	Air blending
Air conditioning	Extra
Interior trim	Cloth seats, nylon headlining
Floor covering	Carpet
Jack	Screw pillar
Jacking points	One at each corner
Windscreen	Laminated
Underbody protection	Bitumastic wax, zinc-coated sills

(Source: *Autocar*)

The high-performance version of the Rover SD1 range, the Vitesse model, was launched in 1982. This publicity photograph differs from previous ones in so far as the car is shown at speed at what is obviously a motor racing circuit, instead of the more usual formal pose, typically outside a stately home.

Vitesse's fuel-injection system was completely different from the US version's fuel injection system, which he feels was a poor design.

The Rover Vitesse was certainly a mean-looking beast. With its big, black polyurethane foam rear spoiler, 15-in spoked alloy wheels with fat Pirelli P6 low-profile tyres and 'Vitesse' name along its side, it could not be mistaken for any other Rover SD1. At the heart of the car was the fuel-injected V8 engine, which developed 190bhp. Its makers claimed that its 0–60mph time of 7.1 seconds was faster than that of any current production saloon then available on the UK market – 'the fastest production Rover ever', with a top speed of 135mph. Maximum power was claimed to be produced at 5,280rpm and a massive 220lb ft torque at 4,000rpm. They also stated it had the added benefit of proven reliability(!), excellent fuel economy and low cost of ownership (fuel consumption was better than in carburetted V8 SD1 cars).

The aerodynamic improvements made to the Vitesse were strictly functional. The large rear spoiler was designed by Rover's own styling department with input from the motor sport department. It was no mere cosmetic addition: at 100mph it added no less than 86.4lb ft downthrust to increase the loading on the rear tyres and improve roadholding and stability. Injection-moulded fairings were attached to the sills at the leading edge of the rear wheel arches (which helped protect the lower bodywork from stone damage), and with its lowered suspension, these aids significantly improved the aerodynamics of the early Vitesse, reducing the drag coefficient to 0.36, compared with 0.405 for the 1982 3500SE SD1 model.

The Vitesse had 10.15-in diameter ventilated front disc brakes with four-piston AP callipers, which had been developed for the Metropolitan Police and then used on the racing cars. It sat 1in lower than other SD1

The name badge denotes this SD1 as a Vitesse – however, with its other distinguishing features you hardly need to refer to the badge.

cars on its race-bred, modified suspension. Spring and damper rates were increased by around 20 per cent in the MacPherson strut front suspension. For the first time on a production Rover, variable rate coil springs were combined with self-levelling Boge Nivomat damper units (which had been uprated and re-calibrated) on the torque tube live axle and trailing link rear suspension. Solid rubber bushes in the Watts linkage improved lateral location of the axle and enhanced stability at high speeds.

The Lucas fuel-injection system fitted to the Vitesse was developed from that fitted to Australian market V8 Rovers for emission control reasons. In its Vitesse form, it was much more powerful – a new air-flow meter and electronic control unit and different distributor advance and retard characteristics were required to adapt the 'L' system from the low-compression 150bhp Australian specification engine to the high-compression Vitesse engine. The engine had a higher compression ratio of 9.75:1, and improved gas flow from reshaped valves with modified stems. The cooling system was uprated to stand up to additional stresses on the engine. The gearbox was strengthened by having shot-peened gears and stronger bearings (like those used in Jaguars) to cope with increased torque. The camshaft was unchanged. In common with the rest of the SD1 range at that time, the power-assisted steering rack ratio was changed to give 3.0 instead of 2.75 turns from lock to lock, and the system was now arranged to reduce the servo effect at speed, with the fitting of a new, falling output pump.

The Vitesse was very well specified, and cost £14,950 on its introduction. Metallic or black paint were no-cost options, but early cars came only in three colours – Monza Red, Silver Leaf and Moonraker Blue. The reason for this and for there being only one interior colour (grey) was to keep production costs down.

Graham Lewis was involved with designing different seats for this car. Early on, the design team bought a pair of Recaro seats and retrimmed them. However, the seats which ended up in the production car were totally redesigned.

Inside, the car had deeply padded, sports-style front seats and head restraints finished in grey velvet. It had a straight grain walnut panel on the instrument facia and matching walnut door fillets. Standard equipment included trip computer, digital clock, stereo radio/cassette player with four speakers and electric aerial, sliding steel sunroof, electrically operated, bronze-tinted windows, central door locking and front and rear inertia reel seat belts. Options included an electric sun roof and air-conditioning, but not leather seats.

A limited slip differential was neither envisaged nor tried out on road-going production Vitesse cars (in spite of company press launch claims that such an option would be offered on all Rovers from spring 1983). The design team felt that there was no requirement for this item. John Davenport tried one out on a road car, but did not believe it made a lot of difference – as he told me, the Vitesse had a lot of traction anyway.

The new Rover Vitesse certainly had the right appearance for a high-performance model. As Rover motor sport driver Tony Pond recalls, if you left a gleaming Vitesse parked somewhere, you would frequently return to the car to find it was surrounded by admirers – in his words, 'the Vitesse has street cred'.

The company deliberately aimed to change the SD1's image with this new high-performance car, from the older company director or businessman, to a different market sector altogether – a younger, upwardly mobile customer. It was billed as the 'elite, executive express'. Trevor Taylor, director of sales and marketing, was quoted as saying:

> We are extremely pleased with the way the sales of the new Rovers have gone ahead in some of the toughest market conditions possible.
>
> The addition of the Vitesse gives added strength to this very fine range of cars and will enhance the name and image of Rover saloons.

Sales brochures boasted of the

> confident, race-bred stance of the fuel-injected Rover Vitesse. With contoured, wraparound front seats faced in Sculptured Plain velvet, and features such as a 3-band electronic stereo radio/cassette player, this superb Rover offers all the luxury you'd expect from the performance overlord of the range. Use that close-ratio gearbox to explore the 190bhp powerhouse, with its potential 0–60 time of just 7.1 seconds, and you'll immediately recognize a rare breed of car.

Rover Vitesse – the Ultimate SD1?

The earlier Vitesse shared the same frontal styling as Vanden Plas models, with the small air dam incorporating twin spot lamps. The 'Vitesse' name decal could be deleted on later cars, as many customers considered this item to be tasteless on such an up-market car. The fairing protected the body from stone chips and aided aerodynamics.

This, then, was a luxury grand touring machine, with all the usual Rover refinements:

> 'race-bred' does not involve any compromise of comfort or furnishing. The Vitesse is exhilarating, certainly; but it also cossets with luxury. It is a totally practical business or family vehicle.

Austin Rover certainly marketed the Vitesse model very successfully – vigorously promoted and advertised, it was the model which everybody wanted. Even members of the Royal Family had them – Prince Edward had a Silver Leaf Vitesse, while Captain Mark Phillips was stopped for speeding by the police in his Moonraker Blue, late twin-plenum Vitesse. John Davenport remembers that there was a big trade in Austin Rover dealers selling Vitesse rear spoilers at the time – one dealer sold one Vitesse car and thirty-seven rear spoilers over the same period of time!

Austin Rover advertisements capitalized on the successes of the cars in racing. An advertisement from July 1981, headed 'Formula for Success', showed a Group 1 racing Rover saloon, and pointed out that Rover won the Tricentrol British Saloon Car

Rover Vitesse compared

	Price (£)	Max speed (mph)	Acceleration 0–60 mph (sec)	Overall mpg
Rover Vitesse	14,950	130	7.6	21.8
Audi 200T	14,313	125	17.5	19.1
BMW 735i (auto)	18,500	129	7.8	19.6
Jaguar XJ6 4.2 (auto)	15,989	127	10.0	16.8
Mercedes-Benz 380SE (auto)	18,800	131	9.1	20.0
Opel Monza S 3.0E	14,591	133	8.5	23.0

(Source: *Autocar*)

Championship at Silverstone on 21 June. It stated:

> There are five outright Rover winners to choose from. See your local dealer for a test drive now.

Later advertisements continued this theme. One, which showed a road-going Vitesse on the racing track, proclaimed:

> There are some places you can take advantage of the Rover Vitesse. Legally.

Another very well known Rover advertisement carried the slogan 'Leader by nature – Paris by lunchtime – car by Rover', showing a Moonraker Blue Vitesse. The makers claimed:

> Its the meanest, fastest and most stylish Rover ever built ... The low, sleek shape is more than just distinctive. It aids and abets both the legendary Rover performance as well as the superb, reassuringly responsive handling ... With luxury and refinements that will soothe and cocoon you, whether you're in town or on the motorway.
>
> The classic Rover Vitesse. Pace-setting, fuel-injected elegance.
>
> For those who have no time to waste.

Unfortunately there were times when the owners of these cars would probably have liked to have changed the slogan to something more appropriate, such as 'Broke down in Dover – never made Paris – car by Rover'!

Needless to say, the motoring press loved the new Vitesse model. *Car* magazine called it 'the poor man's Aston Martin'. Most were impressed by its performance, handling and roadholding qualities, and good fuel economy. *Autocar* managed to obtain a top speed of 133mph, and 0–60mph time of 7.6 seconds, with good overall fuel economy of 21.8mpg, (they believed owners would get 23–25mpg). They concluded:

> It may lack the overall sophistication of some of its peers, yet we ended up liking the car almost for this very reason. It has a distinctly 'animal' character all of its own.

When comparing the Vitesse with the new BMW 528i and Saab 900 Turbo, *What Car?* concluded that:

> for sheer excitement there is nothing to touch the Vitesse. It is a thrilling high-speed racer, a magnificent long-distance mile-eater and a sensibly predictable handler ... in the face of such fabulous flexibility and

such spectacular performance no genuine enthusiast need look anywhere else.

Autosport also liked the new model, commenting:

> Once on the move, the tough V8 emits a delightfully macho wail when howled through the lower gears – music to the ears of sports car enthusiasts or, indeed, anybody with soul – it diminishes to a melodic incidental tune once top is engaged and the Vitesse really displays its mile-eating *vitesse* ...
>
> Overtaking is a joy in the Vitesse as it possesses so much mid-range torque that dramatic intermediate speed progress can be made in any gear ...
>
> So well sorted is the Rover's suspension in stiffened form that the car is as much in its element away from the highways as it is loping along the motorway or dawdling round town where minor jitters over irregularities in the road are the only concession when it comes to ride quality. Front and back seat passengers all commented on the comfort of travel in a variety of conditions.
>
> Although the Rover still rolls noticeably when asked to change direction sharply or swiftly, it does so to a far lesser degree than the older versions in the range. Having said that, the Vitesse corners beautifully consistently and in a neutral attitude even when driving near the very high limits of adhesion offered by the 205/50 VR15 P6s. A touch of understeer is evident if anything while rear end breakaway can only be induced on very slippery surfaces, hooligan antics in low gears or gross mismanagement of braking and cornering lines ...
>
> The brakes are well up to scratch ... pulling the big car up magnificently with vast power ... Stability under braking is one of the Vitesse's strongest suits ... The excellence of the Rover during braking allows deep entry into corners with complete safety and heightens the enjoyment of the car immeasurably.

Like all SD1 cars, the Vitesse could swallow vast quantities of luggage with the rear seat folded down.

The writer concluded:

> The Rover Vitesse is, overall, a fine multi-purpose machine and a driver's car in every sense. Torquey, and ultimately very powerful with no economy penalties as such, the car's able roadholding and tremendous stopping power will endear it to the enthusiast as much as its racy good looks while its luggage capacity of up to 43cu/ft (seats folded) make it double as a capacious family car or businessman's workhorse. At the price you get an awful lot of motor car, well built and eye catchingly packaged ...

The spectre of Solihull unreliability raised its ugly head, however, when *Motor* carried out a 25,000 mile, long-term road test of the Vitesse. They experienced a number of problems, including steering rack failure, brake judder, trip computer not functioning correctly and electrical problems – they also had to replace the radiator. The writer awarded the car two stars for reliability, commenting:

There was a moment, as I stood on the M25 next to a dead Rover, when even two stars seemed generous. The fuel pump relay had given up the ghost.

Motor concluded:

> In a year's time, the Vitesse will have disappeared, and with it the fine V8 engine and great combination of mile-eating performance and sporting character. But it must be hoped its successor is less troublesome.

Changes to the Vitesse were in line with those made to other Rover SD1 models up to 1986. Not everyone liked the Vitesse name on the side of the car, and from early 1983 customers could choose not to have these decals on their car. Between January 1984 and about twelve months later, automatic transmission was available as a no-cost option, but after the Vanden Plas EFi model came out, Austin Rover felt this option was no longer justified. From May 1984, wood trim became burr walnut, replacing the satin-finish type. In October 1984, the model was facelifted, and the car no longer appeared with 'Vitesse' decals and rear wheelarch fairings. The Vitesse model now

Unlike the Vitesse, the Vanden Plas EFi model came with leather seats as standard. Later SD1s from the early 1980s had bright door kickplates showing the Rover name.

The large rear spoiler on all Vitesses was an aerodynamic aid to improve rear tyre grip with down-thrust.

had rubber bump-strips along its sides, and a deeper front spoiler (developed from racing, and unique to the Vitesse model); fog lamps were deleted, as there was no room for these items with the new shape of spoiler. Final external changes to the car came in October 1985, when the car had its radio aerial built into the rear window, and from January 1986 indicator repeaters were added to the car's front wings.

From November 1985, the twin-throttle plenum chamber Vitesse was introduced as a homologation special, to keep the racing cars competitive. These cars were developed by Lotus, who radically altered the car's

Rover Vitesse – the Ultimate SD1?

The V8 engine with enhanced performance through fitting fuel injection and other modifications, as installed in the majority of Vitesse cars (top). Note the large lump on the side of the plenum chamber which houses the single throttle butterfly. Shown below is the engine as developed in conjunction with Lotus for racing purposes – with the twin throttle plenum body and air intake branch casting clearly visible. However, to gain full advantage over the earlier version of this engine, a fiercer camshaft than was supplied with the car is required.

fuel-injection system. Rover built these cars at Cowley as ordinary single-plenum cars, and sent them to Lotus, where they were converted into twin-throttle plenum Vitesses. The cars required numerous modifications, including a different electronic control unit. These cars had two identical throttles of 65.05 internal diameter – earlier Vitesses only had one. All the twin-throttle plenum Vitesses had more massive, forged steel valve rockers (with a manual adjustment screw, welded solid to them, for race homologation purposes. However, the WL9 racing camshaft was not fitted to the production cars because Austin Rover believed that this would have reduced the engine's refinement. The engine was easily distinguished from the earlier one by two intake hoses attached to the plenum chamber. The twin-throttle plenum Vitesse had more torque than the earlier model, and although Rover made no claims in their advertisements (because

Rover Vitesse – the Ultimate SD1?

Both of these cars are examples of very late Vitesses, and have a much larger air dam fitted under the front bumper with no provision for spot lamps – this being a direct result of experiments carried out by the works racing team programme. Also, it can be seen that the rear wheel fairings are no longer fitted, together with the absence of side name decals (which had ceased to be offered) and the addition of side repeater indicator lights to the front wings in common with other SD1 models at that stage.

re-homologation would have been needed), it could produce over 200bhp. In practice, though, owners needed to change the camshaft to a different one, such as the high-lift WL9 camshaft used on the racing cars to obtain a noticeable difference in performance from the earlier Vitesse.

Although the number of twin-throttle plenum cars built has been disputed, British Motor Industry Heritage Trust records suggest that 500 of these cars were produced. John Davenport claims that 500 had to be built in order to get the car homologated for racing purposes. What is not disputed is that the twin-throttle plenum Vitesse are the rarest of all SD1 cars built (even if the Rover SD1 Club is somewhat over-represented by them – over 160 people who own these cars are members)!

The Rover Vitesse was always intended to be a specialist product by its manufacturer, and was never intended to sell in large volumes. Its job as an image-builder for the SD1 range was enhanced by its competition successes and resulting European television coverage, which boosted the sales of other SD1 models. It was only available initially on a built-to-order basis from the factory – dealers did not hold stock of these cars. Nor was the Vitesse available from all Austin Rover dealers.

It was envisaged that around 2,000 of the new cars would be sold between October 1982 and the end of 1983 – two-thirds of these cars were expected to be sold to British buyers, and Germany and Italy were intended to be major markets for the remainder. In the event, just over 1,600 were sold during this period – after great demand in the early life of the model, the number of Vitesses sold actually declined over the years they were produced.

When the last Rover Vitesse car was built in July 1986, 3,897 of this particular model had been produced in total (of which 383 were automatics).

From May 1984, a new high-performance model of the Rover SD1 was introduced. At a price of £15,775 it was, for a few months, the top-of-the-range SD1 model. The Vanden Plas EFi model was basically a more luxurious version of the Vitesse, with the

---

Rover Vanden Plas EFi specification

As for the Vitesse, except for:

Transmission
Type   GM type 180 3-speed automatic, with torque converter

Internal ratios and mph/1,000rpm
- 3rd   1.0: 1/25.0
- 2nd   1.45: 1/17.2
- 1st   2.39: 1/10.5
- Rev   2.090: 1
- Final drive gear   2.85: 1

Equipment
Interior trim   Leather seats (velvet seats a no-cost option), nylon headlining

(Source: *Motor*)

Rover Vitesse – the Ultimate SD1?

Rover Vanden Plas EFi compared				
	Price (£)	Max speed (mph)	Acceleration 0–60 mph (sec)	Overall mpg
Rover Vanden Plas EFi	15,775	128.2	8.9	22.5
Audi 200E	14,944	120.2	12.2	24.4
Jaguar Sovereign 4.2	18,995	128.0	10.5	15.7
Saab 900 Turbo 2.0	13,987	113.5	9.7	20.8
Vauxhall Senator CD 3.0	12,896	127.7	9.9	20.3
Volvo 760 GLE	14,331	113.4	10.3	19.5

(Source: *Motor*)

Vitesse's fuel-injected engine, suspension and brakes, slotted into a Vanden Plas specification bodyshell, but with automatic transmission. Customers had wanted a version of the SD1 with Vitesse performance but with the luxury of a Vanden Plas interior. Harold Musgrove, the company chairman, was seen driving around in one of these cars early on in 1983.

Austin Rover claimed that the new model had a 0–60mph time of 8.2 seconds and a top speed of 130mph. It did not have the rear spoiler of the Vitesse, but did come with the Vitesse alloy wheels. A 'Vanden Plas EFi' rear badge helped to differentiate this particular model. The model kept its original front spoiler from start to finish (unlike the Vitesse cars, which had a new shape of spoiler from 1985).

The Vanden Plas EFi was only ever available in automatic form. It came with leather seats (although box velvet ones could be substituted at no extra cost), and was so well equipped that the only option the customer had to pay for was air-conditioning. Metallic or black paint were no-cost options. The car was aimed at a more conservative buyer than the Vitesse – the businessman or managing director who was more interested in a luxurious interior than in the

Apart from its Vitesse-style wheels, this badge was the only give-away that this car was anything more than an 'ordinary' Vanden Plas.

scintillating performance of the manual Rover Vitesse model. *Motor* described it as 'a car for fast moving executives.'

Austin Rover sales brochures boasted that:

The exclusive Vanden Plas EFi combines supreme elegance with the awe-inspiring power of the fuel-injected V8 engine. And when you sink into seats upholstered in

Connolly leather, with detachable leather seat headrest pads, you'll begin to understand the sheer perfection of this pedigree car. Every Rover refinement is here, from the deep pile footwell rugs to the luxuries of cruise control and an advanced in-car entertainment system. A fine-tuned suspension and effortless performance complete this sporting aristocrat.

Generally, the motoring press preferred the Vitesse model to the automatic Vanden Plas EFi. *What Car?* magazine commented that the car was:

> not as happy a blend of qualities as we had hoped. And it is not such an appealing package as its five-speed Vitesse stablemate which costs a few hundred pounds less.
>
> ... it does not have the easy, relaxed and fabulously flexible feel of the manually geared Vitesse ... the engine often revs needlessly high in response to slight throttle pressure ... it is irritating to be able to hear every change in the engine's note and every beat and pulse in its operation.
>
> ... the auto is less satisfactory ... in the surprisingly sluggish getaway from rest it allows, and in its occasional reluctance to kick down for quick overtaking.
>
> ... But for the luxury owner the least welcome addition to the VDP's specification is the noise – sporting perhaps, but out of keeping with the pleasantly plush trimmings.

Motor liked the EFi model more than *What Car?* journalists had done, although they only managed to obtain a 0–60mph time of 8.9 seconds, and a top speed of 128.2mph, with overall mpg of 20.2 (the automatic transmission sapped a lot of power). They remarked on the poor amount of interior space in a car of this size, and the noticeable amount of wind noise when travelling. They noted:

> From the outside, the eight-year-old SD1 shape still looks modern – confirming that it was well ahead of its time when launched.
>
> ... On the move, it makes the same 'efficient'-sounding induction whuffle as the Vitesse, but is never less than super-smooth and always beautifully responsive to the throttle.
>
> ...Air conditioning is the only option listed for the Vanden Plas – in contrast to the lengthy list provided by certain German car makers ...

Motor concluded:

> Those who value the attractions of a large car powered by a big – but in this case not so lazy – V8 engine, and who value the touch and smell of a traditional 'English' interior will know what the Vanden Plas is all about. The name has lost some of its exclusivity now that it has been applied to volume models in ARG's line up – but at this end of the scale, it still means something special.
>
> It is not the last word in sophisticated engineering, and it is certainly flawed in a number of important areas. But it also has some very positive attributes: it provides more obvious and cosseting luxury than the Vitesse.
>
> Ultimately it is still a prestige car with character – and one that offers more performance than anything at a cheaper level than the V12 Jaguar. As a value-for-money executive express, it will be a hard act for Austin Rover to follow.

The Vanden Plas EFi model was only ever intended to sell in low volumes. In fact, a mere 1,113 models were produced altogether (422 in 1984 and 691 in 1985), which means that this model is one of the rarest of all the SD1 cars. The car was only produced until late in 1985.

12 The Rover SD1 in Motorsport

One of the major surprises of the 1980s was the development of the big Rover SD1 cars into very successful race cars. Work cars were raced between 1980 and 1986. The competition successes were seized upon by Austin Rover's marketing department and promoted to increase sales of the entire Rover SD1 range of cars. Later on Rovers were also rallied, and obtained some good results (especially in 1985). The main features of the racing and rallying, together with the major results, are listed later in this chapter.

How did a large 3.5-litre saloon car become involved in motorsport in the first place? It all started in 1980, when British Saloon Car Championship regulations increased the capacity in the largest class from 3 to 3.5 litres, thus allowing the Rover cars to enter the fray for the very first time. At the time, the company had made good progress in racing and rallying the TR8 cars. John Davenport, director of motorsport throughout the Rover SD1 era, realized that the Rover, with its high-performance 3.5-litre V8 engine and its good handling features, had the potential to be developed into an excellent racing car. He managed to persuade Peter Murrough, one of the directors of Jaguar Rover Triumph, to use the SD1. Not everyone at the company was keen on the idea – Ray Horrocks, who was managing director at British Leyland, was reluctant to race a big-engined large saloon car after the previous failure when Jaguars had been tried out.

In 1979 John Davenport approached David Price Racing and asked them to

The Rover SD1 was first raced in 1980 with cars prepared by David Price Racing. Here, Rex Greenslade was the driver of the Triplex-sponsored 3500 in Round 4 of the British Saloon Car Championship at Silverstone, where he finished in second place.

The Rover SD1 in Motorsport

The fuel-injected Vitesse was raced from 1983, when Steve Soper and Rene Metge won the RAC Tourist Trophy Race. The last time Rover had won this trophy was in 1907.

develop the car for saloon car racing for the 1980 season. The car was tested during the year, and was successful enough to lead to two cars being prepared by David Price (to be driven by Jeff Allam and Rex Greenslade) and raced in 1980. These Group 1 cars produced about 250bhp and had a modicum of success, including a win at Brands Hatch, and the Empire Trophy. Peter Murrough was happy (he could announce these results to dealers trying to sell the SD1 cars) and the company was keen to continue with the racing programme.

There was a feeling that the David Price cars had not been as competitive as they could have been. Tom Walkinshaw Racing (TWR) offered to race the cars for the 1981 season in an attractive deal for the company – for less money than they had paid the previous year to David Price Racing, and the contract was also linked with performance. TWR raced the factory Rovers between 1981 and 1986 very successfully and won numerous races (including several 1-2-3 results)

first in the British Saloon Car Championship and later in European Touring Car events.

In 1981, the Rovers became faster and more competitive, and defeated the Ford Capris on the race track. From 1983, the newly homologated fuel-injected Vitesse was raced, and just about blew everything else off the race track. The cars produced around 290bhp, weighed 2,552lb (1,160kg) and had 30lb more torque than the standard road-going Vitesse. Group A regulations allowed the use of a five-speed gearbox with variable ratios (compared with the SD1 fixed-ratio gearbox). The highlights of this period included an unexpected win by Steve Soper and Rene Metge in wet conditions against Jaguar XJS cars and the BMW coupés in the 1983 RAC Tourist Trophy, only the second outright victory for Rover since 1907. The Rovers dominated the 1983 BSCC, but Rover renounced their claim to this title when a dispute arose. The 1984 BSCC was won by private entrant Andy Rouse in his Rover Vitesse. In 1985 and 1986, it was contracted that Tom Walkinshaw had to be one of the drivers (the idea being to get him more involved, and therefore, hopefully leading to more wins). After the TWR Rovers narrowly lost the ETC in 1985 to the Volvo 240 Turbo cars, Win Percy took the 1986 title, ahead of the BMW 635 CSi models. The 1986 twin-plenum Vitesses had better air flow and breathing, which made a marked improvement on the performance of the race cars. In 1987, private entrant Tim Harvey won Group A of the BSCC in his Rover, but in 1988, Rovers started to disappear from the race circuits.

The story of the dispute which started in June 1983, and dragged on for many months, was an interesting one. Behind it lay personality conflicts and people trying to settle old scores. On 25 June at Donington Park, Edward Grace International, who entered and prepared driver Frank Sytner's BMW, protested about the engine of Allam's Rover and the internal measurements of the rear wheel-arches of all three Rover works cars. As a result of this, an RAC enquiry took place, which went on for months, into the eligibility of the Rovers. It was alleged that the Rovers had breached the regulations by using non-standard valve rockers made from steel and manufactured by Bahco with an additional screw. There were even rumours that the rockers had come from a Volvo, as Volvo rockers had been used on TR8 rally cars. In reality, Tom Walkinshaw had come across two pallets of old rockers at the Solihull factory and found the rockers and valve gear there. No one knew exactly what car they had come from but John Davenport believes they came from an earlier attempt to put the Rover V8 engine into the Land Rover, and had sturdy valve gear. He says that the steel valve rocker arms were standard production car items. Tom Walkinshaw took these rocker forgings and managed to find enough spare material to also make an adjusting screw and nut.

Various claims and counter claims led to the Shawcross Enquiry, which was heard in

Rovers dominated the 1983 BSCC. Here, the Sanyo-sponsored car of Jeff Allam and Hepolite-sponsored car of Steve Soper finished in first and second place at Silverstone, on 2 October.

the High Court. John Davenport believes that the whole matter was counter-productive for TWR and Austin Rover. It cost the two companies a great deal – in June 1984 their expenses were estimated to be £100,000 plus costs (which were, no doubt, substantial). The civil court action was brought to question the validity of the Enquiry, and took up more time and involvement by senior management than the programme justified. The matter was resolved when Austin Rover renounced the 1983 drivers' and manufacturers' titles, and withdrew from the 1984 BSCC. They continued to take part in the European Touring Car Championship, and added a third car to the team. Not surprisingly, the motoring press were very disappointed by the withdrawal of the works Rovers from the BSCC. *Autocar* commented:

> Rovers will still be there of course – in private hands – but it is a shame to see the factory pull out because of politics when they are firmly in front on the track.

Autosport, too, regretted the decision to withdraw:

> The success of the Rovers during the past four seasons has ... been effective in ridding the marque of its 'Auntie' image ... That Austin Rover should feel themselves forced into the position of having to withdraw in mid-season because of a succession of administrative problems with the sport's governing body, the RAC MSA, is most regrettable. But their decision is understandable.

The whole matter had been extremely badly handled by the RAC MSA. *Autosport* remarked:

> The RAC Technical Committee has appeared to be hell-bent on nit-picking within the regulations – not only with the Austin Rover cars – and handing out the most severe and damning punishments for the pettiest of irregularities of the sort that would be put right without question by competitors dealing with more cooperative and understanding scrutineers.

From this point the works Rovers competed in European races, where they achieved many good results.

Rover take on BMW – an oft-repeated sight of the 1980s (in real life, though BMW may have lost a few races, they won the company!)

The Rover SD1 in Motorsport

In 1984, the privately entered, ICS-sponsored Vitesse of Andy Rouse won the BSCC. Here he is seen taking on one of the TWR works-sponsored cars.

In April 1982, Rover revealed their new Group 2 rally car, which had been tested over the winter months by Colin Malkin. This could produce 290bhp and had four twin-choke Weber carburettors. John Davenport gave the project the go-ahead – it was seen as an ideal choice, a large comfortable car to compete in long-distance international rallies, such as the proposed Paris-Dakar event of autumn 1983 (which never took place). The rallying team had gained experience from the TR8 race and rally cars. According to John Davenport, these first SD1 rally cars 'were a bit crude as rally cars went, but worked OK'. The first appearance of the car was at the Esgair Daffydd Rallysprint, with Tony Pond driving. Tony Pond also drove these cars in several Middle East rallies, but they had not been properly tested, and the rallies were not successful. These cars were eventually sold off and reappeared in rallies driven by private entrants.

Interestingly, a number of people within the company had believed that the SD1 would not make a good rally car, because it was felt that it was too large to compete with Ford Escorts and other small fast cars. John Davenport himself believed it was too large. However, Tom Walkinshaw believed it would make a good rally car, and from 1983, TWR built the works rally cars.

One of the highlights of 1983 for Rover motorsport enthusiasts was the annual Austin Rover Rallysprint event, which was held at Donington Park in November. Drivers from the world of Grand Prix motor racing – Nigel Mansell, John Watson, Derek Warwick and Danny Sullivan – competed against rally drivers Stig Blomqvist, Pentti Arikkala, Jimmy McRae and Tony Pond.

Tony Pond was the clear favourite, having won the event for the past three years, and being the only competitor to have driven the Rover rally car previously – however, a tie-break decider meant that he was only just beaten by Nigel Mansell.

It may not have been as fast as his Williams-Renault Formula 1 car, but future champion Nigel Mansell made full use of the power of the V8 engine to help him to win the Austin Rover Rallysprint in 1983.

Tom Walkinshaw driving with John Davenport navigating in the Group A TWR car in 1984. John Davenport was pleasantly surprised by the SD1's rallying abilities.

The Rover SD1 rallying record

winter 81/82	A Group 2 rally car is built by the factory in order to assess the car's suitability as a long-distance rally car.
spring 1982	The first appearance of the rallying factory-built Rover car at the televised Esgair Daffydd Rallysprint, driven by Tony Pond.
Sept 1983	Tom Walkinshaw, partnered by John Davenport, entered their first rally in the car at Bianchi, in Belgium.
Nov 1983	The Austin Rover Rallysprint event at Donington Park was won by Nigel Mansell.
1983	Private rally entrant, Ken Wood, won the Scottish Rally Championship in a Rover rally car.
Aug 1984	Tom Walkinshaw and John Davenport completed their first rally, the Russek Manuals event. In the same month, Tony Pond and Rob Arthur gave Rover their first Group A victory (finishing fifth overall) on the Hunsruck Rally. In the same month, Bob Fowden and Hywel Thomas (private entrants) won the Epynt Stages event in the Welsh Tarmac Championship in a Rover SD1.
Sept 1984	Tony Pond and Rob Arthur dominated the Manx International Rally, and took first place in Group A.
Oct 1984	Private entrants Ken Wood and Peter Brown finished in second place in the Trossachs Rally, taking the Esso Scottish Rally Championship.
1985	Tony Pond and Rob Arthur finished in third place overall, winning the Group A class (for cars over 2,000cc) in the Shell Oils Open Championship, with their Computervision-sponsored Rover Vitesse. Group A wins included the Welsh International, Ulster, Scottish International and Circuit of Ireland Rallies.
Oct 1985	Following their third place in the September Cumbria Rally, Ken Wood and Peter Brown won the Wilsons U-Bix Kingdom Stages Rally in their Rover SD1.

(Source: *Autosport*)

Private entrant Ken Wood had a number of good results in the Rover Vitesse rally car.

The Rover SD1 in Motorsport

Rallying the big SD1 Vitesse with great success in 1985 – Tony Pond and Rob Arthur won the Group A class in the Shell Oils Open Championship in the Computervision-sponsored car.

By 1984, the Tony Pond car won some rally classes and categories; 1985 was the best year, when he won his class in several rallies in the Computervision-sponsored Vitesse, and won the Group A class (over 2000cc) in the Shell Oils Open Championship. The 1985 rally car produced 305bhp and in forest specification weighed about 3,036lb (1,380kg). By the 1986 rally season the Metro 6R4 was ready, so Rover withdrew the works SD1s from rallying (leaving private entrants to compete in them), and Tony Pond drove the Metro instead. Tony Pond believes that the earlier success of the Rover SD1 in rallies led to the development of the 6R4.

The Rover SD1 was never seriously considered as a four-wheel-drive rally car at any time, according to John Davenport. The reason for this was that it would not conveniently convert. A four-wheel-drive Rover 800 was actually completed, (based on the Metro 6R4) but was never actually rallied.

On rally cars, TWR mechanics would change the back axle of cars as a preventative measure – even so, this was the Achilles' heel of the rally car. Quick release brake connections meant that mechanics could change an axle in about ten minutes during an event in cases of failure. As well as the weakness of the crown wheel and pinion, the Watts linkage had been known to collapse under rallying conditions. On a few of the very late rally cars, the Watts linkage was replaced by the use of a more robust Panhard rod for lateral location of the rear axle. The throttle potentiometer on rally cars was changed regularly to ensure reliability. After every rally, each car was stripped down to the body shell and totally rebuilt.

The struts on TWR Group A race and rally cars were specially made up from castings in high tensile alloy, and were offset more than on the standard car to allow for bigger wheels to be fitted. An additional variation from the standard production Vitesse was that the lower ends of the struts were located by fabricated 'A' frame track location arms, rather than relying on the single forged arm located by an anti-roll bar. The TWR rally cars were identical in all respects to the race cars, apart from ride heights and spring and damper rates – rally cars were slightly higher than normal SD1 road cars, and lacked the big front spoiler of the TWR race cars. Both used large ventilated disc brakes with callipers all round, and had a limited slip differential (manufactured by GKN). The cylinder blocks of all motor competition engines had cross-bolted main bearings in order to strengthen them. Race cars had manual steering, while rally cars were usually power-assisted (although there were a few exceptions – cars rallied by Tony Pond at Esgair Daffydd and in the 1985 Welsh Rally did not have this Feature).

A recirculating, forced feed oil supply (with oil cooler) was needed in the rear axles of both race and rally cars, because of the weakness of the crown wheel and pinion. An electrically powered Stuart Warner flexible vane pump was used under the driver's control, and was switched on once the oil had warmed up. On rally cars, twin headlamps (similar to those used on US-specification SD1s) were frequently used as these were much more effective than normal UK production SD1 headlights. Any improvements made to SD1 race and rally cars were usually Tom Walkinshaw's ideas.

The suspension systems of TWR race and rally cars were totally uprated. The rear axle was standard, though toughened. All suspension tie rods and track control arms were fabricated so as to be adjustable in length and pivoted from rose joints, rather than the rubber bushes used in fixed length links as found on the standard production road car. This gave much better location of the wheel assemblies, with the facility to easily make minor adjustments to the suspension geometry in

TWR mechanics carrying out work on the Pond/Arthur rally car. The car (below) is probably having its back axle changed – this was a real weakness on the rally cars. With plenty of practice, mechanics could change one of these items in about ten minutes.

The Rover SD1 in Motorsport

This view of a racing Rover illustrates the impracticality of the massive 'cow catcher'-type air dam for everyday use in speed hump infested British streets.

order to 'set up' the car for different events. This feature is common practice in motorsport – the fact that it gives a harsher ride, of course, being of no consequence. The turrets for the front suspension were moved inwards compared with the standard car, in order to fit larger wheels.

Water-cooled disc brakes were tried out on the race cars – water-cooling was a distinct advantage at Donington and other similar circuits. Adjustable camber was tested on the rear axle of track cars – this was accomplished through divided half-shafts with constant velocity joints located immediately inboard of the hubs.

Tony Pond has some good memories of racing the big Rovers, which he believes represented the pinnacle of saloon car racing at the time, and gave a much-needed boost to Rover's image. In his words, 'it flew the flag and won – it was the last proper racing Rover.' Tony Pond obtained some very good results in the SD1, including a win at Silverstone in 1984. In his opinion the SD1 was a versatile car, and an excellent race car, with stable, well-balanced handling – it was 'well sorted – a nice car to drive – it was a lot of fun driving, and very safe.' The cars were bulky but always competitive. Drivers knew that they had a good chance of winning in these cars. John Davenport remembers the SD1 as a wonderful race car which was beautiful to drive. The only problem was that the race cars did not have power-assisted steering, which meant that steering was very heavy. Tom Walkinshaw maintained that they did not need power-assisted steering on the Rover race cars, but most of his drivers would have disagreed with him about this!

Tony Pond's worst memory of racing the Rover (or indeed, of racing *any* car) was the Spa 24 Hours race in 1984. This took place in appalling driving conditions, in torrential rain, and in the dark. Tony remembers that he could not see a thing – 'It was really hairy.' It was the only time in his racing life that he really did not want to be there. At one point, Tony Pond's car was in the lead, but his car ran out of fuel and did not finish.

The Rover SD1 in Motorsport

Scottish power! The man behind Rover SD1 motor sport success – Tom Walkinshaw, celebrating one of the many Rover victories in ETC races. Win Percy, who can also be seen in the picture, won the driver's title in 1986.

After the first hour of the race, Rovers were 1-2-3. An *Autosport* writer noted that:

> Although they did not win, the sight of three Rover Vitesses leading at one point was heart-warming nevertheless.

All the Group A racing Rovers were sold off after use, if they were not scrapped during races. In March 1987, after the works Rovers were withdrawn from motor racing, three TWR-prepared Group A Vitesses were advertised for sale in *Autosport* magazine.

Tony Pond has some good memories of rallying the Rover SD1, especially of 1985, when he won the Group A class of the Shell Oils Open Championship. Although the Rover was a heavy car, it was competitive

A most unusual advertisement appeared in Autosport *magazine in March 1987. If you could afford them, you too could own three works TWR-prepared Group A Rover Vitesses. Many of the ex-TWR cars were bought and raced by private entrants.*

against smaller, faster cars such as Opel Mantas. Tony Pond says that the SD1 rally car felt like a much smaller car when he was driving it, 'something like a biggish Escort'. Ian Beveridge, who was in charge of SD1 rallying, says that the Rover was relatively easy to drive, but required a great deal of mechanical sympathy, particularly for the rear axle. The crown wheel and pinion was far too small, and even when made from special material had a very limited life. He believes that Tony Pond's ability to judge just how hard he could drive the car without breaking it got Rover the results. Other drivers found it less easy to judge the pace for this large, relatively heavy but powerful car.

In rallies, one of the tricks the team employed was to squirt water and soap on to the tyres to limit traction during rapid starts, thus helping to prolong the life of the transmission. Rover rally cars were driven on the road to rallies, in order to run them in first.

Tony Pond's worst memory of rallying the big Rover was the Circuit of Ireland Rally in April 1985, having broken a toe the week before the rally. The competition department modified the servo brake for him, but driving 'was a very painful experience'. Despite this, he finished in sixth place overall, and won his class. Rover motorsport cars had heavy clutches, but were no different in this respect from their motorsport competitors.

John Davenport has one particularly unpleasant memory of Rover rallying. At the first rally he entered as co-driver for Tom Walkinshaw, the Bianchi Rally in Belgium in September 1983, the car was well placed when it had a puncture and went off the track and hit a tree. John broke a couple of ribs.

After the 1986 season, TWR stopped racing the Rover SD1 cars. After political changes, when Harold Musgrove (chairman), Mark Snowden (managing director) and Peter Reignier (finance director) left the company, things suddenly changed – these three people had been strong supporters of motorsport. John Davenport knew that the new Rover 800, being front-wheel drive and with Honda V6 engine, would be no good for motorsport. Although the company had a contract with Tom Walkinshaw for the forthcoming 1987 season, the people in charge decided not to continue with Rover racing. Tom Walkinshaw was paid off and not held to his contract. John Davenport believes this was a mistake, and thinks that it would have been sensible to continue racing the SD1, and that it would have carried on winning (as privately entered Rovers continued to do in the 1987 BSCC).

Tony Pond, seen here relaxing between stages of the Welsh International Rally in 1985 – this particular year was the best one for SD1 rallying.

The Rover SD1 racing record

1980   A change in the regulations for cars in the British Saloon Car Championship (increasing the engine capacity limit from 3 litres to 3.5 litres), led to the first Rover SD1 cars being raced, with preparation carried out by David Price Racing. Early successes included a win by Jeff Allam at the British Grand Prix supporter, and a 1–2 for Rover at Donington, with F1 champion Alan Jones driving the winning Rover.

1981   From this season, the works Rovers were prepared by Tom Walkinshaw Racing. This season, Peter Lovett and Jeff Allam came first and second in the 2,501–3,500cc class in the BSCC, ending the Ford Capri's long reign at the top. Rover won six out of eleven rounds, (four of these being 1–2 victories).

1982   Rover took six outright victories in the BSCC, compared with five for the Capris. The class title went to Rover's Jeff Allam at Silverstone in October, after a tie-break decider. Rene Metge won the French Saloon Car Championship outright in a Rover.

1983   The fuel injected Vitesse was raced for the first time this season. Rover dominated the BSCC, winning every round they entered, with Steve Soper winning the Championship, and several results were 1–2–3 for Rover. Rover also won the manufacturers' title for the first time ever. In September, Steve Soper and Rene Metge won the RAC Tourist Trophy outright, (the first time Rover had won this trophy since 1907). However, an enquiry which started in June marred the season. For the first time, the Rover works cars entered the European Touring Car Championship, and at the Spa 24 Hours race, Allam, Soper and Lovett took a worthy third place.

1984   This season started well for Rover, but was marred by the above ongoing dispute. Tony Pond won at Silverstone, and Rovers took the first four places at Thruxton. After the Shawcross Enquiry, the dispute was resolved in June – TWR renounced the claim to the 1983 title, and the TWR Rovers were withdrawn from the British Saloon Car Championship in the middle of the season although they continued to compete in European Touring Car races. The British Championship was won by private entrant Andy Rouse in his Rover Vitesse (prepared by his own company, with help from Austin Rover Motorsport)

1985   The TWR Rovers had a number of successes, including 1–2–3 results at Monza and Donington, and wins at Nogaro and Vallelunga. In spite of Walkinshaw and Percy winning at Jamara, Rover narrowly lost the ETC driver's crown to Volvo.

1986   The twin-plenum Vitesses were raced for the first time. At the BSCC Brands Hatch Grand Prix supporter in July (at the time of the launch of the new Rover 800), Jeff Allam won and Rovers took five out of the top six places! In the European Touring Cars Championship, there were wins at Monza, Donington and the RAC Tourist Trophy race at Silverstone. Estoril was the final race for the TWR works Rovers – Win Percy and Tom Walkinshaw came second, which meant that Percy won the drivers' title and Touring Car Championship from the BMWs by one point.

1987   Private entrants have successes in the British Touring Car Championship driving Rover Vitesses. Tim Harvey won three races (Oulton Park, Silverstone and Brands Hatch), compared with Dennis Leach's one win – Harvey won Class A in the BTCC from Leach by one point.

1988   Dennis Leach was the sole entrant in a Rover Vitesse in the BTCC – however, he withdrew two-thirds of the way into the season.

(Source: *Autosport*)

A sight to warm the heart of any Rover SD1 enthusiast – the large red and white Bastos-sponsored Vitesse ahead of BMWs (and Volvos) in a European Touring Cars race.

13 The Police Cars

The V8-engined predecessor of the Rover SD1, the Rover 3500 P6B, was a common sight on British roads during the 1970s. It was liked by the police, particularly for motorway patrol duty. British Leyland were naturally keen that police forces should maintain their brand loyalty to the company, and police forces throughout the country were given the opportunity of test-driving the new Rover SD1 as the various engined variants of the car were introduced from 1976 onwards.

Thames Valley Police liked the SD1's design – they praised the steering, brakes and load-carrying versatility, but thought the boot sill was too high (the designers had deliberately made the boot sill high to give the car strength – it made the car stronger in an accident). Other forces, such as Wiltshire Police, were also impressed by the 3500, but were deterred from buying it by its high price. Eventually, Wiltshire Police only had two 2600s, which they allocated to their driving school for instruction to advanced level, keeping these vehicles for three years.

A road test of the Rover SD1 appeared in the *Police Review* in 1976. The police road-tester was extremely impressed by the new Rover. He commented on the car's attractive styling and appearance. The Rover's rack and pinion steering was pleasant to use, and its roadholding and handling were excellent. The ride was good, the brakes were efficient and the seats were comfortable. The car also had ample room for carrying police equipment. The road-tester remarked on the 'odd-looking' non-circular steering wheel on early SD1 cars. The only faults he found after testing the car for hundreds of miles were the position of the handbrake (too far away from the driver) and the restricted rear view due to the steeply sloping shape of the rear window. Nevertheless, the road-tester remarked that it was one of the finest cars he had driven for a long time, and recommended it for patrol work, crime squads and for use as Q cars.

Many police forces, including Thames Valley Police, replaced their high-speed Jaguars and BMWs with new Rovers. The company was proud that police forces throughout the UK bought the new Rover for high-mileage patrol vehicles, and advertised the fact in sales brochures. One sales brochure stated that the West Yorkshire Police Force ordered a new generation of SD1s for motorway use after being so impressed with their previous SD1 vehicles, which had completed 100,000 trouble-free miles.

For these reasons the Rover SD1 was a popular police vehicle, seen everywhere throughout Britain in the familiar 'jam sandwich' livery during the 1970s and 1980s. The combination of a versatile hatchback body and impressive performance (particularly of the 3500 version) made the car a firm favourite with many police forces. It was feared by speeding motorists everywhere! The V8-engined version was definitely most popular with police forces. Northamptonshire Police used both 3500s

The Police Cars

West Yorkshire Police used a number of manual 3500s for patrol duty, and found they were generally reliable. It can be noted that much of the equipment on the roof was mounted on a detachable rack to minimize marks and holes in the bodywork, together with the badge and stripes (probably being self-adhesive transfers) – both of which could be easily removed, thus maximizing the residual value of the vehicle on completion of its police service.

and 2600s between 1976 and 1982, and found the 3500s to be the more reliable of the two models. Cleveland Constabulary used six-cylinder SD1s for three years, but found they were a continual problem in mechanical terms and suffered from corrosion – these cars were not well liked by that force, which was glad when they were disposed of. A few forces did not use SD1s as police cars – Surrey Police, for example, had only one SD1, which was the Chief Constable's official vehicle.

West Yorkshire Police only had two 2600 vehicles for staff car purposes – the rest of their police SD1s were manual 3500s. Police Sergeant Clarke drove these 3500s during his time as traffic patrol officer. He found them to be generally reliable, powerful and spacious vehicles, and believed they were superior to the previous Rovers (P6Bs) which they had used. However, the SD1 was not without problems – early models suffered from excessive body corrosion. The gearboxes were prone to bending the reverse arm when subjected to high reversing speeds (which could arise in particular situations where they needed to reverse out of somewhere in a hurry). The cylinder-head gaskets on a significant number of vehicles needed to be replaced repeatedly.

Generally, police SD1s were of basic specification with no sunroof, no central locking, no electric windows, non-power-assisted steering and plain cloth seats. They had rubber mats in the boot and frequently had Minilite aluminium wheels fitted. Contrary to popular belief, police cars were not very different to standard, but they tended to have improved brakes and specially strengthened 5-speed gearboxes. Typically, the interior trim was rather basic, there was an accurately calibrated speedometer, a high-output alternator, a large capacity battery, and the wiring was modified to permit installation of radio and other specialized police equipment. The main modification to the bodywork was the police 'jam sandwich' livery of white with a red reflective strip,

Rear view showing the very large illuminated police 'stop' sign styled in the form of a very large spoiler. This car was a 3500SE model used by West Mercia Police.

The Police Cars

The Rover SD1 was chosen by the Department of Transport to use in their anti-drink-driving campaign poster, being a typical police car of this era (copyright of the Department of Transport).

although different forces had their own liveries. Many forces removed the front spoilers on Series II cars, which were prone to damage even in non-police use. Police SD1s were kept for several years, and experienced extremely high mileage. One late 3500 used by Cheshire Constabulary Patrol Fleet reached 204,000 miles (a lot of miles, even by police standards).

John Davenport's motorsport department got involved in sorting out problems for police forces, and other similar tasks. In the early 1980s, the Thames Valley Police at Kidlington, Oxfordshire had problems with their V8-engined SD1s, which were used for motorway patrol duty. The suspension on these cars tended to bottom out on the road with all the police gear carried in the back

The Police Cars

and with standard SD1 dampers and springs fitted. At Cowley, uprated springs and dampers were put on to the car, and an anti-roll bar was fitted to the rear. After these modifications, the car worked well, and the police were pleased – they sent for a further four kits to improve other SD1 police cars.

The Rover SD1 police car was even used on a poster as part of a Department of Transport anti-drink-driving campaign during the 1980s.

The Metropolitan Police Rover SD1s, commonly seen all over the London area during the 1970s and 1980s, are worth examining in some detail. The story of how the Rover SD1 was developed and tested for police use by the Metropolitan Police is a fascinating one. I was fortunate to discuss this subject with Richard French of the Metropolitan Police. He worked as the chief test driver in the Test and Development branch of the force during a period of almost six years when the Met developed the SD1 as a police car.

The Home Office dictated the Metropolitan Police's vehicle requirements, and required them to buy British cars. The Triumph 2500s and Rover P6Bs (3500s) needed to be replaced by similarly powerful cars of the same engine sizes. The cars had to be automatic with large engines with high torque and low stress. (Hard driving in London traffic is tiring in a manual car, and manual cars are more expensive to run, with clutch wear and so on.) The cars also needed to be large, four-door family cars. The only suitable cars were Rovers (2300, 2600 and 3500) and the Ford Granada, which was eliminated from the list because it was German-built. The Rover 2300 was discounted because it lacked performance. Finally, the Rover 2600 was chosen, on cost grounds – it was about £600 cheaper than the 3500, and this meant considerable savings on fleets of cars. The Metropolitan Police actually argued in favour of the Rover 3500. They believed that it was a much better engineered car than the 2600, having the same engine as the P6B which had proved to be reliable, and the engines shared similar parts. In Richard French's words, though, 'The Met were lumbered with the 2600.'

This picture shows an early Rover SD1 in comparison with the then current P6 3500S, which was about to be replaced as a traffic patrol car by the Metropolitan Police. The new 3500 was in experimental livery – it was the first time that the force had used a striped livery on their cars.

The Police Cars

The Metropolitan Police first tested the Rover SD1 and evaluated it for police use from 1978 onwards. It was tested for reliability, and modified in conjunction with the manufacturer. The Met had an experimental 2600 pre-production car. This particular car was a Rover 3500 with a 2600 engine, and was very fast. The Metropolitan Police spent about six years developing and testing various Rover SD1 cars for police use. The SD1 worked well for the police in the end.

One of the reasons that convinced me to buy a Rover 3500 Vanden Plas in 1986 was the fact that the cars were so popular with police forces throughout the UK – after all, police cars have to be reliable, above anything else. However, there were a number of reasons for this. First of all, the police made some modifications of their own. Secondly, they regularly serviced the cars every 6,000 miles (more frequently than recommended by the manufacturer). Finally, police cars were not fitted with various electrical items, such as sunroofs, electric windows, central locking and other accessories, that were prone to fail on ordinary SD1s.

The Metropolitan Police started off by using Rover 2600s as area cars and as traffic patrol cars. The specifications for both cars were as follows:

- Area car – crime car, assigned to robberies, murders and so on. Markings used, single blue light on roof. Radio aerial behind, amber and yellow reflective stripes. Police coat of arms. Police radio and siren. Each car carried in the boot six road traffic cones and two accident signs.

The very first Met traffic patrol car, a 2600 model. This had twin spot lamps on the roof and no police signs on the front or rear of the car.

The Police Cars

These pictures show the first Met traffic patrol car revised to a later police specification, with twin blue lamps on the roof and police signs on the front and rear of the car.

The Police Cars

The Met area cars were all 2600 models. These slightly later cars can be differentiated from the traffic patrol cars by the single lamp on the roof. Note that these cars now had Minilite alloy wheels fitted, to dissipate the heat caused by hard police driving techniques.

Small fire extinguisher kept under front passenger seat. On top of the roof, an aerial identification mark (for identification by police helicopters).
- Traffic patrol car – assigned to serious accidents, speed limit enforcement and so on. Markings used, two blue lights on roof, police sign on front. Illuminated police 'stop' sign on rear. Aerials and markings the same as area cars. Specially calibrated speedometer for monitoring speeding. Radio and siren. Public address system (there was a loudspeaker under the front bumper). Each car carried in the boot six road traffic cones and two accident signs, two long torches with blue flashing lights to put on road cones (used to warn of accidents), first aid kit, two foam fire extinguishers, a blanket and a broom to sweep up accident debris. On top of the roof, an aerial identification mark.

The Metropolitan Police also used Complaints cars in traffic. These were unmarked cars with sirens and a magnetic blue lamp to put on the roof and a calibrated speedometer. Royalty Protection cars were unmarked cars with magnetic police signs and a magnetic blue lamp to go on the roof. All police SD1s had hard-wearing plain cloth seats, and stripped interiors compared with standard SD1s. Traffic patrol cars were used for 18–20 hours, and area cars for 24 hours at a stretch in normal police use.

All area cars were Rover 2600s. The first traffic cars were Rover 2600s also. However, after running these for about twelve months, it was found that they lacked power and acceleration. Therefore, from about 'V' registration Rover 3500s were used and from then all traffic cars were Rover V8s. The Metropolitan Police kept the cars for about 89–90,000 miles or 8–10 years.

The Metropolitan Police made a number of modifications to their Rovers. The main problem with early SD1s was with the brakes. Police driving in London traffic puts a strain on standard brakes (police forces elsewhere did not suffer from this problem – it was entirely due to the particular problems encountered when driving in London traffic conditions). The early disc brakes failed in tests. The police, together with the manufacturer, made a number of modifications, including fitting non-asbestos brake pads and thicker rear brake drum castings. Minilite alloy wheels were fitted to dissipate heat (the Rover alloy wheels wouldn't fit after these modifications had been carried out). The Master cylinder was changed, and the car was transformed. The problem of water getting into the master cylinder reservoir was cured by fitting a blanking grommet to cover the end of the level float pin. Some of these improvements to the braking system, prompted by the Metropolitan Police, were eventually adopted by Rover for the SD1 range in general (the ventilated disc brakes being reserved for the high-performance Vitesse and Vanden Plas EFi models).

- Suspension: this had to be uprated on all SD1s, because of heavy loads carried and the hard driving conditions the cars were subjected to. Stronger springs and dampers were fitted. Front and rear bushes wore out rapidly, but this problem was overcome by changing them regularly.
- Rear axle: this tended to wear out rapidly, owing to heavy driving. The problem was overcome by changing it regularly.
- Anti-roll bars: there were problems with these items breaking. When kerbs were clouted the ends sheared off where they meet the track control arm. This problem was overcome by regular inspection and changing them for new ones.
- Transmission: transmission problems were no worse than in the P6B. The auto-

The Police Cars

All the later Met traffic patrol cars were 3500s, as the police found the 2600s lacking in power for their particular needs. The V8-engined cars also had proven reliability.

matic transmissions were built in conjunction with the manufacturers and were upgraded for hard police use. The average general usage was around 42,000 miles before they wore out. The majority of SD1s, including Vitesses, were automatic (although there were a few very late, manual SD1s).
- Engines: standard engines were used, (the police did not modify or improve the performance, contrary to common belief). The Metropolitan Police had no problem with the V8-engined SD1s, which rarely went wrong, proving very reliable.

The V8 engine had worked well in the P6B, and did also in the SD1. However, the 2600s were another matter altogether. There were cylinder head problems on the six-cylinder cars – problems with the setting up of the valve clearance after overhaul because of the instability and flimsiness of the camshaft carrier casting. Another problem was camshaft lubrication failure leading to overheating and distortion and failure of the cylinder head. The Metropolitan Police overcame this problem by keeping a stock of overhauled cylinder heads and continually changing them. History has an unhappy habit of repeating itself – the Wolseley Six-Eighty of the 1950s, which was popular with police forces then, also suffered from cylinder head problems.
- Servicing: brakes were checked at 2,500 miles before they were modified by the manufacturer. After this, brakes lasted

6,000 miles. Brake pads were changed every 6,000 miles and discs were changed every 12,000 miles (more regularly than the manufacturer recommended on later cars). All SD1s (early and late ones) were serviced every 6,000 miles. The engine oil was changed every 6,000 miles (the 12,000-mile interval recommended by the manufacturer on later cars leads to sludging).

- Steering: all police SD1s had non-power-assisted steering even the 3500s. This, I am reliably informed, made steering extremely heavy. The police made a modification to tighten up the steering column height adjust. The original non-circular SD1 steering wheel was thrown away, and a round one fitted. However, when the company bought out the later steering wheel on Series II cars the police kept these.
- Heater: the problem here was with the heater matrix leaking. As any SD1 owner who has carried out this task knows, it is a swine of a job to replace this item. (Fortunately, with plenty of practice, the police mechanics got this job down to a fine art!)
- Electrical system: as the police used basic cars, with manual window winders, and without standard SD1 electrical items, they did not experience the level of electrical problems suffered by ordinary SD1 owners. Police SD1s had a specially calibrated, large mechanical police speedometer. There were no problems with the fuel injection systems of the few Vitesses.

For some reason the 2600s wore out throttle cables at an excessive rate. The Metropolitan Police relied on the manufacturer for crash testing and safety tests, as they have limited resources. Bodywork problems on the SD1 were no worse than normal – all police vehicles suffer from body fatigue, owing to the hard driving conditions they endure.

To improve ground clearance, the spoilers on Series II SD1s were removed and thrown away. After doing this, there were no problems with speed humps and so on, as commonly experienced by ordinary SD1 (especially Vitesse) owners. The end caps on the plastic bumpers were always falling off and getting lost on the road – they had to be continually replaced.

SD1 ministerial cars, were just standard Rover SD1s. No Rover SD1s were ever armoured or reinforced by the police. The SD1 was not suitable for armouring, which would have been either too expensive or impossible, owing to the hatchback shape of the car and the pronounced curve of the window glasses, which would have had to be replaced by bullet-proof glass. The body was also too light for this to be carried out. Jaguars were armoured – as heavier saloon cars they are more suitable for this treatment. The Rover SD1's ministerial predecessor, the P5B, had been armoured, as it was a heavier car. (It is rumoured that the purchaser of one second-hand P5B wondered why his car lacked performance and guzzled petrol excessively – eventually he found panels of boiler-plate under the carpet!)

There were two types of Royalty Protection SD1s: marked cars with police signs, and unmarked cars. Royalty Protection cars were used as special escort cars, to escort members of the Royal Family being driven in their own vehicles. Very late Royalty Protection cars were Vitesses.

Only one Rover SD1 was specially built at the request of the Metropolitan Police. They required a high-speed, unmarked car. This was a red, manual 3500 which was specially tuned by the police, and which had a Holley carburettor fitted. However, the car was still not fast enough for the police's liking. Therefore, the Met approached Janspeed, who improved the performance of this vehicle for them.

The Janspeed police SD1 had twin turbochargers blowing through Weber carburet-

The Police Cars

tors, and was intercooled. This car was very fast, had a top speed of 160mph, and did 0–60mph in 5 seconds. The engine was not bored out. The Janspeed car was used on motorways and was in service for some time. The Traffic Police had requested the car as an unmarked car for high-speed pursuit, chasing supercars and so on. The police eventually decided that the public's unawareness of a powerful unmarked car bearing down on them at high speed was inherently hazardous, and therefore it was decided to discard the concept. The Met then converted it back to a normal 3500, and it was eventually sold off as an ex-police car in the usual way.

The Metropolitan Police kept the SD1s for a long time, eventually replacing them with the Rover 800 Series cars and Ford Sierras. At the time of the introduction of the Rover 800, which had not been fully tested by the police, the company had large stocks of SD1s hanging around in the showroooms. The Metropolitan Police offered to buy all these cars as a special job lot, at a substantial price discount, just to keep the police fleet going. These cars included Vanden Plas models, manual 2600s and Vitesses. It is rumoured that the 2600 Vanden Plas cars were bought for only £1,200 each. The cars were resprayed white, but had ordinary SD1 features including sunroofs and electric windows.

An interesting point is that soon after the police were seen driving around in the Vanden Plas models, members of the public complained about policemen travelling around in luxury, wasting public money. These cars were taken back to the workshop to have their Vanden Plas badges replaced by ordinary ones.

The Vitesses were only purchased by the Met in the last job lot of SD1s they bought, as these models were usually expensive. They never bought these cars on their own, but in the last job lot at a substantial discount. Like other police SD1s the Vitesses were factory-standard cars, and not modified in any way. It would have been too costly both in time and money to carry out performance improvements. The rear spoiler on the Vitesse was incorporated into the police sign.

The Rover SD1 was the last major car project developed with a car manufacturer by the Metropolitan Police. In this respect, then it marked the end of an era.

A pair of old SD1s being used for driver training on the Hendon skid pad by the Metropolitan Police, and displaying some minor bodywork modifications. The car nearest the camera shows large, extended mud flaps to minimize the effects of spray from the lubricant used on the training area, and the rather strange replacement of working tail-lights with some sort of mock-up, combined with the use of straps to secure the tail-gate.

14 Overseas Variations and Other Unusual SD1 Cars

The product planning department wanted a much wider market for the Rover SD1 than just the UK, and intended it to sell well world-wide. The North American market (ironically, as it turned out) was identified as one where the SD1 cars could obtain a good market share.

The Rover SD1 sold well in a number of European countries, especially Italy and Germany. Over 30,000 Rover SD1s were sold in Europe from the car's launch in 1976 up to the end of 1981. In 1981 alone, total sales increased to over 6,000 in the nine major European markets, with 1,700 sold in Italy by August 1982, year a 100 per cent improvement over Italian sales for the previous year.

Generally, differences between UK and overseas market cars were minor. The 1983 year Vitesse model sold in Switzerland had a carburetted engine, owing to legislative restrictions here. Not all models were available in every market – for example, the 2000S model was only sold in Italy, where the tax regime meant that models of 2.3 litre size or below were favoured by many buyers – this model had a better trim specification than the UK market 2000. The 2000 model was intended to increase Rover sales in Europe by between 25 and 30 per cent and was sold in Italy, France Belgium and Holland. The 2300S model was sold in Austria, France, Belgium and Holland, and the 2600S Model was sold in Portugal, Spain, Switzerland, France, Belgium, Germany, Italy and Holland. From 1982, top-of-the-range Vanden Plas models were sold in Germany, Italy, Holland, Belgium, Switzerland, Austria and Spain.

As well as Europe, Rover SD1 cars were sold in the Far East, Middle East and Africa. However, there were three overseas versions of the SD1 which had significant differences from the UK-specification cars – these were for North America, Australia and South Africa.

Millions of pounds were spent developing the North American version of the Rover SD1, to comply with US emissions regulations. Rovers for this particular market were made in the UK and exported.

The North American cars were very similar to the V8-S model, with air-conditioning, alloy wheels and cross-ribbed velvet upholstery. A sunroof was optional, but there were no rear headrests. The car had a smaller, three-spoke steering wheel, similar to that used on the Triumph TR8. Obvious external differences from UK cars included huge, impact-absorbing black bumpers, four round headlamps and small Union Jack badges fitted to the front wings. A manual five-speed gearbox was standard (with automatic an optional extra). The emission-controlled Lucas fuel-injection system had feed-back oxygen sensors, and the car was fitted with two catalytic converters and a different exhaust manifold. Its V8 engine had a compression ratio of 8.13:1 and produced 133bhp at 5,000rpm. In road tests, the manual car

Overseas Variations and Other Unusual SD1 Cars

Testing the Rover 3500 for the North American market were, on the left, Peter Allinson of Rover Triumph, and right, test driver Dan Archer. The photograph shows the 50,000 mile exhaust emission test vehicle – on the car's bonnet are the North American emission regulations.

had a top speed of 116mph and a 0–60mph time of 10 seconds.

The Rover SD1 got enthusiastic reviews by the American motoring press. However, a number of circumstances ensured that the car was a sales failure in America – the scene of Rover marketing disasters both before and after the SD1. For one thing, the 1979 oil crisis raised oil prices dramatically, and US legislation of the time did no favours to large-engined cars. In addition, the pound was rising in value in relation to the dollar. When the Rover was launched in 1980, the American public had started to buy smaller cars, and V8- engined cars were somewhat out of fashion at this point. The changes made to the car to suit US regulations, with its twin round headlamps and thick black bumpers, did nothing to improve the appearance of the Bache-styled design – a *Road and Track* motor journalist liked the body styling apart from the round headlamps, commenting that 'you should blame the US government and not Rover for that'. The federalized version was emasculated, compared with UK or European specification cars – John Davenport recalls driving a manual, North American specification car and remembers that its performance was really awful – 'it would not go any faster than

Overseas Variations and Other Unusual SD1 Cars

80mph, was totally gutless and drank petrol'. The North American SD1s suffered from the usual quality control and reliability problems, and lacked good dealer spares and servicing back-up. In fact, *Car* magazine wrote in 1986 that:

> One suspects that if the Rover name did mean anything to Americans, it would probably be 'trouble' – such is the reputation of the Rover SD1.

Another problem in certain overseas markets such as North America and Japan, according to John Bacchus (who was director of business and product strategy at the time), was that a large, expensive hatchback was not acceptable to many people – they wanted a saloon (which could be driven by a chauffeur). A hatchback was seen as 'van-like' and possibly slightly down-market at this time.

The Rover SD1 was launched in North America in June 1980, but was withdrawn during 1981. A mere 1,254 cars were sold (480 in 1980 and 774 in 1981), although some left-overs were sold in later years, after the company had stopped exporting them here. Most of the cars sold in 1981 were 1980 models – only about twenty or thirty 1981 models got to the USA. The North American fiasco is something that those involved would prefer to forget. Mike Cook worked for Jaguar Rover Triumph in North America, and remembers trying to sell the Rover SD1 there. He recalls that British Leyland were broke, and had no funds to advertise or promote the car. Its press launch was combined with that of the Triumph TR8, to save money. The quality of the cars was poor, and they were expensive ($15,900 at the time of launch – by 1981 dealers were offering $2,000 discounts). No real effort was put into selling them. Not surprisingly, the SD1 project was doomed in the US, and dealers were stuck with cars they could not sell.

The Australian version of the Rover SD1 was also modified to comply with emissions regulations, and like the US, the 3500 model only was sold here. It was launched in Australia in October 1978, with a low-compression V8 engine (of 8.13:1 compression ratio), which could produce 136bhp. It came with an Alpinair air-conditioning system which was fitted by the importer, Leyland Australia, after the cars arrived there. The Australian SD1s had automatic transmission, electric windows and alloy wheels as standard. The early model had Zenith-Stromberg carburettors with automatic choke, an air-injection pump and an exhaust recirculation system. The fascia was different from UK-specification cars, and side intrusion bars were fitted in the doors to meet local regulations.

From the middle of 1981, the Australian SD1 became a 3500SE, powered by a 142bhp engine with a Lucas fuel-injection system (which produced less power than UK-specification fuel-injected SD1s) and carrying badges stating 'Fuel Injection' above the V8 badges on the front wings. The engine included an exhaust gas recirculation valve. Unlike the US-specification cars, the Australian fuel-injected cars did not have catalytic converters. The high-pressure fuel pump on these fuel-injected cars was prone to failure. Series II 3500SE models appeared in 1982, and Vanden Plas cars from 1983 – the Vanden Plas model came with leather upholstery, sunroof and carpet mats as standard. Some very late Vanden Plas cars were fitted with Vitesse four-pot front disc brakes. The later fuel-injected cars were smoother and more powerful than the earlier cars.

The Australian *Modern Motor* magazine was very enthusiastic about the Rover SD1 when they road tested it :

> … good cars come fairly regularly these days. Great cars – and make no mistake

this Rover is one – are somewhat rarer to come by.

The 2600 models were not available in Australia through Rover dealers, but a number of these cars were privately imported. Only a small number of five-speed cars were ever imported into the country. Sales of Australian SD1s stopped in January 1986, when tighter emissions control regulations were introduced (it was uneconomic for Rover to meet any further regulations). Spen King can remember being in Australia, and being appalled by the quality of new Rover SD1 cars he saw there – he was horrified by the dreadful fit of the panels. Rover SD1 cars are very rare now in Australia.

With regard to other SD1s 'down under', 1,368 CKD cars were assembled in Nelson, New Zealand between 1979 and 1983 with some local content (such as wheels, tyres, wiring harness and so on). These Rovers were 3500 and 2600 models (automatic and manual cars). The full range of Rover SD1 cars was not available here, although some Vitesses entered the country as private imports.

The South African SD1 was another overseas variant. There were two versions – the V8-engined version, and a 2,623cc six-cylinder engined car. The six-cylinder engine was completely different from the UK six-cylinder engine. This particular unit was an Australian-designed one, derived from the 'E' series Austin Princess 2.2-litre engine and used in the Australian P76 model (a nondescript-looking Leyland barge). After the P76 debacle, Leyland decided to shut the programme down and to transfer the power unit tooling to South African for SD1 production. These cars were assembled in South Africa from CKD kits sent out there from the UK, and had to have over 60 per cent local content to meet import regulations and avoid high import duties – engines, interiors, air-conditioning units and other parts were made locally. Some of the cars were sent to the UK for evaluation purposes. The South African SD1 was codenamed 'Lisa 10' (the Lisa referred to Leyland South Africa). The body tools for these cars were Japanese, and the cars proved to be of much better quality than the UK-produced cars (with Pressed Steel Fisher tooling).

The South African V8-engined models were mechanically similar to UK cars. However, the five-speed model was called the SDS, while the automatic was called the SDE. The two six-cylinder models were called the SD (which had basic specification) and the SDX (a better specified 2600). The different badges and bump-strips on the sides of these cars meant that you could distinguish them from UK SD1s. The early cars came with air-conditioning and electric windows as standard, and without a sunroof. Seat trims were different from UK-specification cars – V8-engined cars had the choice of leather or wool seats, while six-cylinder cars had ribbed cloth material. The back seat on all cars had built in rear headrests. The six-cylinder car produced 110bhp at 4,750rpm and had a compression ratio of 8.75:1. It had similar performance to the UK 2000 model, with a top speed of 101mph and a 0–60mph time of 14 seconds. An owner of one of these cars believes it has better torque and is more economical than a UK 2600 model.

The first cars were assembled at the Blackheath plant near Cape Town in 1977. From 1982, the cars were similar to UK Series II cars, and the Vanden Plas model became available – however, South African cars always had stainless steel bumpers with rubber faces (like 1981 UK V8–engined models). The South African plant ceased assembly of Rover SD1 cars in 1983, when Leyland South Africa folded up. Rover Group figures show that about 14,376 cars were assembled in South Africa. After the project finished, John

Overseas Variations and Other Unusual SD1 Cars

The BMC-derived six-cylinder engine, as used in the South African SD1. This engine was usually found under the bonnets of top-of-the-range Austin Princess cars, and as such, would normally be of 2.2-litres capacity and transversely mounted on top of a sump containing the front-wheel drive transmission.

Bacchus wanted to bring the superior body tooling over from South Africa, but the costs would have been prohibitive.

The story of the Indian SD1, the Standard 2000, is a very unhappy one. John Bacchus was closely involved with this project, which came about because Standard Motor Products Ltd of Madras had the licence to build cars in the early 1980s, although they were only building vans and trucks at the time. John Bacchus believed that the Austin Montego would be manufactured here at some later date and that it would be the car to replace the outdated Hundustan Ambassador. The Standard 2000 idea was well received by the Indian government, and it was envisaged that 3,000 cars would be produced each year, eventually going up to 5,000. Production would get under way while the SD1 was still in production in the UK. By early 1988, about 30 per cent of SD1 body dies had been sent to India.

During 1984, Standard Motor Products of India first assembled the Standard 2000, from CKD kits sent from the UK. The car was actually launched in India on 14 Janu-

Overseas Variations and Other Unusual SD1 Cars

The badges of South African SD1s differentiated them from UK-specification cars.

The vehicle identification plate of this car clearly shows that it was built in South Africa, and is a Rover SDX model. This plate was mounted under the bonnet – on UK-built cars, it was mounted on the driver's side 'B' post.

ary 1985 – the British Motor Industry Heritage Trust at Gaydon have a trophy in their collection which commemorates the event. The Standard 2000 was intended to be a prestigious vehicle in the Indian market, and was expensive, costing about twice the price of a Hindustan Ambassador, at 212,000 rupees (about £12,000). However, the price of the car was no problem – there were sufficient wealthy people in India. It was bought by rich businessmen and government bodies – one good customer was Vice President Desai (who later became the President of India). At least one car was sent over to the UK for evaluation. Rover intended that the Indian factory would supply replacement panels and other spare parts for Rover SD1 cars all over the world, if the quality reached an acceptable standard, at some future date, but this idea was abandoned when the project failed in 1988.

About 60 per cent of the car was made locally, including the engine, gearbox, interior parts and air-conditioning units. The car had an Indian-developed 2-litre engine, four-speed all-synchromesh gearbox and rear axle. The roots of the engine went back to the Standard Vanguard. This four-cylinder, wet-liner ohv unit was completely reworked, and given twin SU carburettors, while cylinder head and pistons were reworked to a Heron head design (similar in principle to that of the Rover P6 2000 engine). The 1,991cc engine had a compression ratio of 8.0:1, and produced 85bhp. The suspension was raised by one inch by putting alloy blocks between the body mounts and the springs – spring rates were changed, and there were no self-levelling struts. The car had air conditioning, electric windows and a comfortably trimmed interior. The fascia was the same as a UK-specification SD1, although locally-sourced instruments were fitted. Externally, it looked like a Rover 2000 model, but had rear badges stating 'Standard 2000', with a traditional Standard-Triumph shield badge on the front bonnet. Graham Robson visited India in 1986 and drove a two-year-old car, which he found remarkably free of rattles.

Rover Group figures show that 3,408 Indian SD1s were built between 1984 and 1986 (432 in 1984, 1,824 in 1985, and 1,152 in 1986), and more had been made by the time production stopped late in 1988. Various circumstances led to the stopping of production. Although the car passed stringent Indian fuel efficiency requirements when it was launched (and thus greatly reduced the amount of import duties to be paid), early in 1988 the company was investigated by government officials – claims had been made that fuel efficiency requirements were being broken by these cars. This investigation lasted for many months and cost the company huge sums of money. In addition, there were labour problems. Eventually, the vehicles were retested and found to be fuel efficient; however, by this time the damage had been done to the company, and it was far too late for the Standard 2000. The company failed, and although some wealthy individuals tried to restart the project, none succeeded.

One interesting variant of the Rover SD1 which, unfortunately, never reached the production stage, was the estate car (its Triumph predecessor had been available as an estate car). Alan Edis was involved with product planning for the SD1 cars, and remembers the estate car project very well. Only two of these prototypes were ever made, in 1976 and 1977, according to British Motor Industry Heritage Trust records. One of these cars was used by chairman Michael Edwardes, the other one was believed to have been used by Prince Charles for a while. Product planning saw the opportunity to extend the SD1 range and it was seen as an attractive option at the

The SD1 estate car – the car in the photograph was one of two prototypes to be built in order to evaluate the concept of the SD1 as an estate car. This particular example spent some time in the hands of the (then) company chairman, Michael Edwardes, for use as his own personal transport.

Overseas Variations and Other Unusual SD1 Cars

Rear view showing the slightly raised roof line to give maximum interior height (above) and a view from the side showing the interior with its rear seat as found in a normal saloon (below).

time. However, the huge cost of investment that would have been needed made the project difficult to justify, as far as management were concerned. In any case, the normal hatchback version of the car offered virtual estate-car carrying capacity.

Alan Edis remembers that enthusiasm for the new prototype was high at first, but too much had to be done to put it into production. Also, the loading height in the back was not felt to be adequate, and potential rivals such a Ford and Peugeot offered larger-capacity vehicles. However, a number of people involved with the SD1 project now believe that British Leyland missed a good opportunity, and that this car, if it had been put into production, could have sold well against the likes of Ford, Volvo and other manufacturers, who dominated this market sector.

The prototype estate car which was used by Michael Edwardes for a while is now part of the British Motor Industry Heritage Trust's collection at Gaydon. Being a 'P' registration car, this vehicle is a standard 3500 automatic, with normal early SD1 interior. The only apparent differences from the normal SD1 hatchback are the square back end of the car, and the rearward-facing fitted seat in the estate boot of the car (probably a child's seat, only intended for occasional use – this seat would be folded down for normal use as an estate car). The other prototype estate car is now in the Haynes Museum in Sparkford, Somerset.

Another variant of the Rover SD1 which was made, although without any factory involvement by the manufacturer, was a convertible. This particular car, a left-hand drive 3500 automatic (in white with a brown interior) was made in 1979. British Motor Industry Trust records prove that British Leyland had no involvement in the conversion of this vehicle – this was carried out by SMC Engineering of Bristol in 1980. All of the pillars ('A' 'B' and 'C' posts) and the roof cant rails were left intact to preserve the structural integrity of the body (which meant that all the window frames were kept from the original saloon – like those on a Citroen 2CV, for example). This convertible was sold again to a Rover enthusiast in 1993, and found its way to Belgium.

With regard to stretched limousine conversions, it would appear that the hatchback shape and lightweight body of the Rover SD1 were not suitable for this sort of conversion. Coleman Milne of Bolton, who are perhaps the best-known coachbuilders of stretched limousines, did not convert any Rover SD1 cars (although they have converted some of the more recent Rover 800 Series saloons). There was a really hideous conversion marketed by the Dove Group, which converted the car into a seven-seater, described by a motor journalist as 'A front-runner in the ugly-duckling-of-the-year contest'. The conversion kit cost £475 including VAT, and consisted of a roof rack, an extension pod, seats on a frame, and a step for the back bumper. The pod was fitted in the rear window frame after the glass had been removed – a sort of dormer window. This whole conversion took the motorist an estimated two hours to complete. When fitted, there was an ugly box shape sticking out of the back of the car with windows in the sides.

Many Rover SD1 customers, especially of the early Series I cars, were unhappy with the spartan interiors of their vehicles. It was not surprising, then, that demand led to specialist companies like Wood and Pickett of London (who were well-known for their customized Minis and Range Rovers) supplying completely fitted interiors, featuring wood and leather. The most popular options were the replacement of the Rover's original plastic instrument panel with a full-width fascia (using existing instrumentation) set in deep burr walnut veneer, wood cappings on all four doors and a four-spoke, polished steering wheel. These items were very expensive – if numerous items were required, the cost was nearly as much as (if not more than) a brand new Rover car. In view of the high cost, very few SD1s received the Wood and Pickett treatment – it is believed that only five of these vehicles are still in existence.

Prices for Wood and Pickett items for Rover SD1 (1980)	
Exterior	
Front air dam	£166.75
Rear spoiler	£109.25
Pair foglights recessed in to airdam	£55.78
Paint highlighting on lower panels	£345.00
Coachbuilders finelining	£63.25
Chromed wire wheels/Dunlop 185/70 VR 14 tyres	£1,259.25
Interior	
Deep burr walnut veneer fascia using existing instrumentation and including matching door capping	£644.00
Polished wood or black leather four spoke steering wheel	£97.75
C type Recaro front seats and complete Draylon retrim of interior	£3,248.75
Head restraints with stereo speakers (Pair)	£212.75
Electric moonshine roof and vinyl surround	£730.25
Mechanical	
Digital keyless entry system	£800.00
Electronic cruise control	£280.00
On board fuel and speed computer	£220.00
Turbocharger (Janspeed)	£1,144.25
Total	£9,377.03
Cost of Rover 2.6 automatic	£8,637.98
FINAL TOTAL	£18,015.01

*Note: prices included fitting and VAT

(Source: *Autocar*)

Many owners have improved the performance of their Rover SD1s, or fitted body-styling kits to their cars. Care le Gant of London offered a range of spoilers and body-styling kits. Other specialists have supplied a vast range of items to improve the performance of the V8 engine – from different camshafts, to turbochargers and bored-out V8 engines. Some people have uprated the performance of the Rover 2600, although items to improve the performance of this engine are much rarer than for the V8. Owners of V8-engined cars can find plenty of information about uprating them in numerous books and magazines.

One contemporary Rover SD1 performance enhancement was available from Janspeed of Salisbury, Wiltshire. This company produced a twin turbocharger conversion, for both Rover 2600 and 3500 models. The original twin turbocharger conversion operated on a draw-through system utilizing the standard SU carburettors (obviously with a conversion of this nature, it was not possible to use an intercooler). However, by the early 1980s, the conversion was uprated to a blow-through

Overseas Variations and Other Unusual SD1 Cars

The Wood and Pickett enhancement of the Rover SD1 interior consisted mainly of items of cabinetwork in the form of door cappings and a fascia panel, the latter being proved with a hinged, glove-box lid. The whole ensemble was mounted on top of the original Rover fascia mouldings and housed the standard Rover instrument and switch assembly. From the photograph, it can be seen that the top of this assembly was rather high in relation to the sight-line of the driver; even if it did not add to visibility problems, it may have induced a feeling of claustrophobia.

conversion utilizing a Weber carburettor, which resulted in a far more driveable and tractable conversion. Towards the end of the production life of the Rover SD1, the conversion was updated again, and had an inter-cooler fitted with standard Stromberg or SU carburettors retained.

Unlike the Rover 3500, the engine of the 2600 had to have its compression ratio reduced on the Janspeed conversion in

Overseas Variations and Other Unusual SD1 Cars

order that sufficient boost pressure could be used. On the six-cylinder car, the compression ratio was lowered by a decompression plate and a longer timing belt was provided to enable belt tension adjustment to be made in the normal way.

The early Janspeed V8-engined car was claimed to have an increase in power and torque of 30 per cent or more over the original car, and it cost about £1,000 in 1979 (as a fit-it-yourself kit, although Janspeed would fit it for you for an extra £120). When *Hot Car* magazine tested a Janspeed-converted 3500 model in 1981 (one of the early Janspeed conversions utilizing the standard SU carburettors), they found the car was very fast indeed, accelerating from 0–60mph in 6.6 seconds, and with a top speed of 138mph. The Janspeed car had plenty of torque, to whisk 'you up hill and down dale in a manner more befitting an Inter-City express train'.

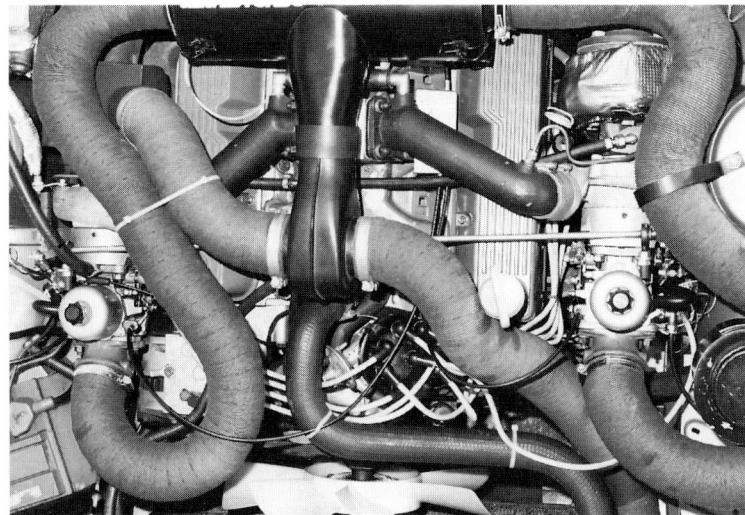

The original Janspeed twin turbocharger conversion installed on a V8 engine used the standard SU carburettors. Almost the forerunner to the later Vitesse model, this particular car carried side decals, rear spoiler and different alloy wheels from the standard 3500 SD1.

15 The End of the Line

From 1979, when BL forged links with the Japanese car-maker Honda, the writing was on the wall for SD1. In 1982, a joint project, codenamed XX, was announced, which would replace the SD1. Even before this, the design team had been considering updating the SD1 with a restyled body and improved running gear.

It was no coincidence that the Rover 800, particularly in hatchback form, resembled the SD1 model it replaced. Roy Axe, who was director of design at Rover at the time of the 800 model, remembers that the SD1's styling influenced both the front of the new car, and also the rear shape of the later hatchback version of the 800. He recalled that the Rover 800 (like the SD1) did not have a front grille, mainly because of the company's disastrous experience with the 'funny face' Vanden Plas model of the Austin Allegro. Gordon Sked, who was also involved with the 800's exterior styling, said that the SD1 was a difficult act to follow:

> ... it is still quite a handsome car, beautifully proportioned.

The new Rover 800 was a very different car from its SD1 predecessor. It was front-wheel

Although the hatchback version of the Rover 800 deliberately resembled the SD1, it was more angular-looking and far less distinctive than its curvaceous predecessor.

The End of the Line

> **Rover SD1 – Main production changes 1976–1986**
>
> Jun 1976 SD1 launched in 3500 form with improved, 155bhp version of the V8 engine already used in Rover P5B and P6B models.
>
> Oct 1977 Six-cylinder 2300 and 2600 versions launched with new Triumph-developed engine. (Low-spec 2300 entry level model was not actually available until May 1978).
>
> Jun 1979 Luxury V8-S version appeared with metallic paint, alloy wheels, leather upholstery and other frills.
>
> Oct 1980 Rover strived to improve the car's flagging image by introducing higher spec 2300S, 2600S 3500SE and Vanden Plas models. Fifth gear ratio on manual V8 raised to 29.7mph per 1,000rpm, making it the highest-geared production car in the world.
>
> Jan 1982 Whole range facelifted with new flush-fitting headlamps, plastic bumpers, different badges etc. and extensively restyled interior and dash.
> New 2000 model introduced with 1,994cc four-cylinder engine developed from the 'O' series unit used in the Morris Ital.
>
> Apr 1982 2400SD Turbo model launched at the Turin Motor Show with Italian VM turbo diesel engine. Mostly aimed at continental market but some sold in Britain.
>
> Oct 1982 High-performance Rover Vitesse launched, with Lucas electronic fuel injection giving 190bhp and 0–60mph time of 7.1 seconds. Vitesse was lower and had uprated, stiffer suspension with four-pot front disc brakes and alloy wheels.
>
> May 1984 Vitesse engine slotted into Vanden Plas-spec bodyshell to create Vanden Plas EFi, with luxury interior and automatic transmission.
>
> Nov 1985 Twin-plenum Vitesse introduced as a homologation special to keep race cars competitive; roadgoing version had slightly better torque than standard Vitesse but no race camshaft, so performance increase was marginal.
>
> Jul 1986 Production of SD1 ceased with the introduction of the Honda-based Rover 800. Around 300,000 SD1 cars of all types had been built.
>
> (Source: *Classic Cars*)

drive, with either a 16-valve twin-ohc four-cylinder, 1,994cc or a 24-valve ohc V6-cylinder, 2,494cc, 170bhp engines and was initially only available as a four-door saloon. For customers used to the torque and performance of the SD1, particularly in V8 form, there was no comparison between the two models. Many motoring journalists seemed to prefer the SD1 – they criticized the 800's very soft suspension (which could make travellers feel car-sick), overlight steering, and lack of low-speed torque, which meant that a lot of gear-changing was needed to obtain reasonable performance.

Some magazines such as *Autocar* carried out long-term tests, and found that poor build quality and electrical and other problems had not finished with the SD1 – one journalist described it as living through a twelve-month long episode of *Tales of the Unexpected*! The styling was considered bland – *Car* commented that the 800 lacked 'presence and distinction ... nobody looked twice at the car ... [[it] won't be talked about in years to come – unlike the SD1'. The 800 Series Vitesse also had mixed reviews. *Performance Car* remarked:

> it's a very different animal to the old rear-wheel-drive Vitesse – in fact it's a lot less animal altogether. It's a far cry from the thundering, booming extrovert roadburner that the SD1 Vitesse so unashamedly was.

Autosport wryly noted that 'we can't see Tom Walkinshaw racing the 800 either ...'

People who worked for the company remember that it had originally been intended to sell the SD1 until well into 1987. Changes in the company, when Graham Day took over as chairman, brought about the early demise of the SD1. When a number of directors left, including Harold Musgrove, the new people in charge made the deliberate decision to break all connections with the old regime. Unfortunately, the SD1 was considered to be part of the old regime, and it was decided to stop its production. Some people, including John Davenport, feel that the decision to drop the car at this point was unfortunate – the SD1 had started to get a better reputation by this time, especially in Europe. The new Rover 800 was a totally different car – many people preferred a large rear-wheel drive car to the front-wheel drive alternative (BMW and Mercedes cars have sold well because of this fact).

At the time that the Rover SD1 was ending its production life, *Car* magazine paid it the compliment of referring to it as:

> the first good car to emerge from the Leyland alliance which nearly extinguished the native British car industry in the '70s.

The Rover SD1 stopped production in July 1986. The very last car to be built was a twin-plenum Vitesse, painted in Silver

---

Late Rover SD1 paint and trim colours

Exterior Colours

White Diamond
Champagne Beige (pale beige)
Targa Red (bright red)
Clove Brown (mid-brown)
Nightwatch Blue (dark blue)
Black
Silver Leaf (metallic)
Cashmere Gold (metallic)
Oporto Red (mid-red metallic)
Silk Green (pale green metallic)
Azure Blue (pale blue metallic)
Moonraker Blue (mid-blue metallic)

Trim

Bitter Chocolate or Claret
Bitter Chocolate
Flint, Caramel, Osprey or Flint/Osprey
Caramel
Prussian Blue, Ice Blue or Flint
Caramel, Claret, Osprey or Flint/Osprey
Flint, Claret, Bitter Chocolate, Osprey or Flint/Osprey
Bitter Chocolate
Caramel or Bitter Chocolate
Flint or Bitter Chocolate
Prussian Blue, Ice Blue or Flint
Prussian Blue, Flint Ice Blue, Osprey or Flint/Osprey

Note: not all colours were available on every Rover SD1 model – for example, the Vitesse had only four colour choices (Targa Red, Black, Silver Leaf and Moonraker Blue). Black and metallic colours were available as a no-cost option on Rover 3500 Vanden Plas EFi models, but were optional at extra cost on all other models.

(Source: Austin Rover colour and trim guide)

The End of the Line

The very last SD1 car to be built in July 1986 was a Silver Leaf Vitesse like the one seen here with earlier Rovers.

Leaf. It was given the registration number of D537PUK, and the Vehicle Identification Number of SARRREWZ7CM345831. This particular car is now housed as part of the collection of vehicles at the British Motor Industry Heritage Trust at Gaydon in Warwickshire.

Looking back with the benefit of hindsight, Spen King feels that the SD1 story is a very sad one – that the car was a good design, and potentially should have been very successful if its development had been carried out properly. He believes that more money should have been spent by Leyland on improving quality, and that less emphasis should have been placed on churning out the car in vast numbers.

Mike Lewis, too, believes that the SD1 is a good design, being 'a practical, good-looking car which is fun to drive on difficult roads and comfortable to be in for long journeys'. He would still consider owning one, if it were not for the very narrow roads in the part of Cornwall where he now lives.

Ultimately, far fewer SD1s were sold than originally envisaged by the company. It was intended that the SD1 would outsell the combined total numbers of Rover P6 and Triumph 2000 models. In the event fewer SD1s were sold than either of its predecessors; a

Early 1970s SD1 sales estimates

Forecast volumes (000s)

Model	1976	1977	1978	1979	1980	1981	1982
2300	3.7	1.0	21.3	20.7	19.8	20.4	19.6
2600	3.7	17.4	18.2	18.4	17.6	18.4	17.4
V8	8.7	13.3	13.9	13.9	13.4	13.7	13.1

(Source: Rover Group)

total of 303,345 Rover SD1s were produced, compared with 329,066 Rover P6s and 316,962 Triumph 2000/2500s. Thus, in terms of production volumes, the Rover SD1 was not considered a success by its manufacturer.

Nigel Heslop remembers that in the early 1970s, the product planning department had envisaged that around 103,000 SD1s would be sold each year. It had been intended that these figures would be achieved by selling the car to Rover P6 and Triumph 2000 owners, and also by making substantial inroads into competitors' market shares – mainly Granada, Volvo and Peugeot. Clearly, this aim was never achieved.

The Rover SD1 was important for three reasons. First of all, it inspired two other car models. The first of these was the Vauxhall Cavalier hatchback of the 1980s, which stylistically looks like a baby SD1. Indeed, I can remember a Cavalier-owner raving about how the new Vauxhall looked like 'the big Rover'. The Cavalier, with its engine sizes between 1300cc and 2000cc, was very much the SD2. The second influence of the SD1 was that it persuaded Ford to introduce a hatchback version of their best-selling Granada model in the 1980s – it was fairly obvious that this model was aimed at people who may have previously purchased Rover SD1 cars. Finally, the Rover SD1 was the last proper, truly British Rover car model before the link with Honda, and eventual ownership by the German BMW car-maker.

The SD1 was a good concept, but was let down by being poorly executed in practice. Problems within the British Leyland organization meant that the car was poorly built and suffered from reliability problems. The Rover SD1, whilst being the best executive car of its era was never made properly. If the Rover had been built to the same standards as its Mercedes and BMW competitors, its story could have been so very different.

The SD2 that Leyland didn't build but GM did – the 1980s Vauxhall Cavalier hatchback showed the influence of the Rover SD1 in its styling.

Rover SD1 Production Figures 1976–1986

	3500 incl. VDP (not EFi)	2600 incl. VDP (etc.)	2300 incl. S	2000	2400 SD Turbo	V8-S	Vitesse	VDP EFi	Total less CKD	Figures incl. CKD
1976	8,654								8,654	8,738**
1977	23,797	2,611	1						26,409	26,537
1978	24,194	23,977	6,184						54,355	54,462
1979	18,565	20,994	5,498			1,040			46,097	46,599
1980	6,521*	12,797	5,896			*			25,214	25,619
1981	8,393	15,803	6,753	1,050	25				32,024	35,885
1982	5,283	8,611	5,582	7,887	4,192		202		31,757	32,885
1983	5,937	10,448	5,453	5,881	3,770		1,437		32,926	33,455
1984	3,354	7,035	3,293	2,909	1,836		1,098	422	19,947	20,379
1985	3,139	5,216	3,037	2,827	258		748	691	15,916	n/a
1986	79	1,080	1,299				412		2,870	n/a
Total	107,916	108,572	42,996	20,554	10,081	(1,040)*	3,897‡	1,113	296,169	(303,345)***

* A separate V8-S figure for 1980 is not available. V8-S models are counted with the remainder of 3500 models in 1980.
** includes production in 1975
*** achieved by adding 1985 and 1986 figures less CKD to annual figures, 1976–84 include CKD.
‡ of which 363 (1983–85) were automatics.

Production at Solihull (1976–82)   192,707

CKD production (1977–83)   14,376

(The fact that this figure is larger than the difference between the two total columns suggests that some, but not all, CKD cars have been counted in column 1)

A small number of bodyshells were built up into competition cars after production had stopped.

(Source: British Motor Industry Heritage Trust)

When Ford decided to rebody their large car in the 1980s, it was decided to follow the five-door format of the Rover SD1 as it was seen as a sales success. The model shown here is a Granada Ghia.

16 Rover SD1 – the Poor Man's Ferrari?

Austin Rover claimed in some of their early Vitesse advertisements that its 0–60mph time of 7.1 seconds was faster than that of the famous Ferrari Dino (however, they omitted to mention that the Dino was the cheapest and least desirable Ferrari of its era, with a mere six-cylinder engine!).

It is interesting to look at the Rover SD1 and compare it more closely with another famous Ferrari – the Daytona (or Ferrari 365 GTB/4, to give it its model number). When first examined, apart from certain styling similarities, the two cars might appear to have few features in common. However, they have more in common than you would think, especially if the Daytona is compared with the Rover Vitesse – the high performance version of the SD1.

The build quality of the Daytona, according to contemporary road tests, was excellent – paintwork and fit of body panels were of a very high standard. Great attention was paid to detail – for example, the stereo speakers were cleverly hidden behind a stainless steel grille. Of course, each Daytona body was hand-built, whereas the Rover was mass produced on the production line. As the Ferrari cost a vast sum of money to purchase in the first place, the customer would expect (and demand) a well-finished car. The less said about the Rover in this respect, the better. For reasons I have referred to earlier on in the book, the build quality and paintwork of the Rover SD1 was poor, especially when compared with similarly priced cars from other manufacturers. This was commented on time after time, both by motor journalists carrying out road tests, and in letters written by members of the public to motoring magazines.

Another difference concerns the sophisticated features of the Ferrari, and their absence in the Rover. The Ferrari had a complicated V12, four overhead camshaft engine with six twin-choke Weber carburettors. The Rover's V8, fuel-injected overhead valve engine was ordinary by comparison – a totally conventional design, almost old-fashioned in some respects. The Daytona had its gearbox at the rear of the car, integral with the final drive, to optimize weight distribution. The Vitesse had independent front suspension, but a live rear axle. The Ferrari had all-round independent suspension of an elaborate nature. It also had pop-up headlights – the Rover did not share this feature.

The Rover SD1 was only ever available in five-door hatchback form (although there was a choice of different engine sizes). It could hold five people. The Ferrari was a two-seater sports car. However, the owner had the choice of buying either a coupé or a convertible version of this car.

Turning, then, to the similarities between these two cars, it is necessary, first of all, to look at the styling features. As we have noted previously, David Bache, the stylist of the Rover SD1, was influenced by the shape of the Ferrari Daytona – the front nose of the Ferrari, with its indicator lights, and

171

the styling crease along the sides of the Italian car, can be seen clearly on the British design. Indeed, the overall lines of both cars are not dissimilar. The Rover SD1 even had swept-up sides, just like the Ferrari. The Daytona lines were completely smooth. There were no door handles or wing mirrors to break up its sleek surface. Although the Rover had these items, it nevertheless had a smooth and clean appearance. David Bache was not too perturbed by comments on the resemblance between the two cars. He told Graham Robson (in *The Rover Story*):

> if people want to call a Rover a four-door Ferrari that has to be good for sales and the image. I'm not complaining.

A common modification to the Rover SD1 is for owners to fit five-spoke, cast alloy, Ferrari-like wheels. These make Rovers look even more like Daytonas than they do already.

Both the Ferrari Daytona and Rover Vitesse were extremely good, high-performance cars (although obviously the Daytona, with its larger, 4390cc engine, had a superior performance of 0–60mph in 5.4 seconds and 174mph top speed (compared with 7.1 seconds and 135mph for the Rover). Both were front-engined, rear-wheel drive cars with masses of torque. Each car had a heavy clutch, which could seem almost lorry-like at times. Other common features were good handling and roadholding (although the Rover had a tendency to understeer slightly), a firm ride, and a large amount of boot space (particularly in the Rover, which took vast quantities of luggage when the rear seat was folded down). The Daytona had a limited-slip differential, which the Rover lacked – therefore the British car, with its enormous

The real thing – the Ferrari Daytona was much better built than the mass-produced Rover SD1; it was also relatively far more expensive to buy. The Ferrari was more reliable than the British car, but had far heavier steering.

torque, could be easily provoked into wheelspin. The steering on the Daytona was not power-assisted, and was incredibly heavy, whilst the power-assisted Rover had light steering (however, when the engine is switched off – thus being without the aid of power-assistance, the Rover is a heavy beast to manoeuvre!) Both cars had electric windows as standard, and had comfortable seats for the driver, offering good support. Both cars were comfortable to travel in for long journeys, and can be called grand tourers. In both cars it was difficult to judge extremities because of the sloping front nose. However, the Daytona owner had a good all-round view out of the rear window. The Rover was extremely difficult to reverse in, because of the lack of vision at the back of the car. Testimony to this fact are the almost obligatory dents to Rover rear bumpers, which are particularly obvious on the early SD1 models.

According to various road tests, both the Rover and Ferrari Daytona were enjoyable cars to drive. Each car's engine produced its own special sound – the Rover had the glorious waffling sound of a V8 engine, which is music to most ears when the car is pushed hard and the Ferrari V12 had its own unique growl. These cars had their own distinctive natures. The Daytona, according to its owners, is a highly individual car – while the Vitesse has a distinctive 'animal' character all of its own.

Both cars have strong engines if these are serviced and the oil changed regularly. Either car can suffer from corrosion problems. Having a steel body, the Italian car suffers from rust in some areas, particularly the inner sills, rear wheel arches and the valance at the back of the car. Rover outer panels and body seams generally tend to corrode – problem areas include the rear wheel arches (worse than the front ones), the bottom edges of the rear doors, edges of the bonnet, at the back of the car around the hatch and taillights and the sunroof. With regard to long-lasting qualities and reliability, the Rover seems to have copied the Ferrari reputation. The suspension parts of the Daytona, the bushes and so on, tend to wear very quickly. The Rover Vitesse shares this feature.

The Rover Vitesse is noted for being temperamental like a Ferrari. Road tests listed a whole range of problems, especially electrical ones. However, according to Daytona owners, the Ferrari is generally mechanically reliable. In a car magazine test of a recent model of Ferrari, the car spent most of its time broken down. This is much more Rover SD1-like behaviour!

The Ferrari Daytona and the Rover Vitesse were both derived from the racing experience of their manufacturers. The Daytona won many competitions between 1969 and 1979. In Group Four of the GT group, they dominated this class at Le Mans in 1972, in the twenty-four-hour race. As we have noted previously, the works Rover Vitesses, prepared by Tom Walkinshaw, dominated saloon car racing during the 1980s. The Daytona had a short production run, from 1969 to 1973. The Rover SD1, in its Vitesse form, also had a short production run, from 1982 to 1986. Although around 300,000 SD1 models of all types were produced in total, only 3,897 Vitesse models were made. This compares with 1,400 Daytonas produced altogether. Only 500 of the late twin-plenum Vitesse were made, compared with 125 of the rarest Daytona model, the Spyder.

The Ferrari and Rover both share reasonable fuel consumption, considering their engine sizes. The Rover Vitesse is especially good in this respect, averaging well over 20 mpg. The Daytona is heavier than the British car. With regard to prices, Daytona parts and servicing costs are expensive – which is to be expected. The car is an excellent investment, and some have sold for between £50,000 and £300,000 (for an immaculate Spyder) in

Rover SD1 – the Poor Man's Ferrari?

The Robin Hood Daytona, for those who wanted something even more like the Ferrari, but who couldn't afford the real thing.

recent years. Whilst not in the same league, many Rover Vitesse parts are quite expensive (exhaust systems, back-axles, spoilers and so on). A Rover Vitesse is not a car to run when you have a limited budget. However, compared to the Daytona, it is a bargain to purchase – ranging from £1,200 to £4,000 (depending on condition), according to the *Classic Cars* magazine price list.

The Daytona is considered by many Ferrari enthusiasts to be the best of the Ferraris (before the mid-engined types and the V6 Dino). Again, the Rover SD1 can be seen as the last truly 'British' Rover – the last completely British-designed Rover before the Honda link was forged. The Rover SD1, especially in Vitesse form, can be considered to be the 'poor man's Ferrari'.

It is perhaps no coincidence, then, that a Daytona replica, based on SD1 bodyshell and mechanical parts, was produced during the 1980s. The Robin Hood Daytona was the first Ferrari Daytona replica to be made. It was produced by Robin Hood Engineering, based in the Nottingham area. Richard Stewart decided to build a facsimile of the Ferrari Daytona 365 Spyder at a reasonable price, and realized that, given the car's styling lines, a conversion of the Rover SD1 was a viable proposition.

Clever modifications to the Rover exterior changed it into an open Daytona, the Spyder. Customers had to provide their own SD1 for conversion. The conversion from four-door Rover to two-door Daytona took around 12 weeks and was a major operation. Mechanical modifications were restricted to the rear axle and suspension, partly to remove the Rover's tail-up attitude. The bodywork was reworked from top to bottom.

The chassis/frame unit was lengthened at the front by around ten inches, keeping the original bulkhead and floorpan. Also, the rear axle was relocated to obtain the same wheelbase as the Ferrari, and the petrol tank was moved into the boot. Adjustable shock absorbers were used to help lower the rear end. Although the SD1 inner panels were retained, all external sheet metal was new, the only visible part remaining from the original car being the windscreen frame.

The conversion needed additional strengthening and re-engineering – having lost the rear doors by grafting on new sheet metal, and having extended the boot area and removed the rear seats (which created space to store the convertible hood) – reinforcing panels were welded into the shell and around the door apertures.

Robin Hood Engineering did not carry out engine modifications themselves, but customers could have their engines tuned by specialist firms – the ordinary Rover V8 or fuel-injected engine could be tuned to produce between 200 and 300bhp. However, in performance terms it could not hope to compete with the 352bhp at 7,500rpm output produced by the real Ferrari Daytona.

The interior had Rover velour seats, but the customer could order leather ones for an extra £1,000. Without leather seats, the total cost of the conversion was £8,500 plus VAT, a fraction of the cost of a Ferrari.

When tested by *Kitcars and Specials* magazine in November 1984, the Robin Hood Daytona was found to be 'firm but comfortable' in its ride and handling, with no 'trace of shake or shimmy when traversing bumpy B-roads'. The road-tester summed up by saying that the Robin Hood Daytona Spyder 'is an impressive and desirable machine and without doubt excellent value for money'.

These cars suffered from corrosion, like their SD1 donors, even though they had been treated with Waxoyl rust-proofing treatment. Robin Hood Engineering did not carry out this conversion for very long – it is believed they stopped making the Robin Hood Daytona after being threatened with litigation from Ferrari.

17 Living with the Rover SD1

What should the potential Rover SD1 owner look for when buying a car? To start with, it is advisable to look for the best example you can possibly afford. Basket cases with rotten bodywork and iffy histories are best avoided at all costs, as you may end up spending your life savings on restoration work.

Which particular model of car should you choose? This is entirely up to the individual, as each type has its own advantages and disadvantages. My own particular preference is for a very late V8-engined model of some type – the later cars have the plusher interior, with wood and sometimes leather seats. If you do long motorway journeys, then the V8-engined car is the one to go for, with the 2600 model a second choice on grounds of engine noise. Most V8-engined cars are automatics, so it is more difficult to find a good manual car. The V8 engines are basically generally reliable, and are good for over 100,000 miles, although the automatic chokes fitted to later cars can play up. The fuel-injection system of Vitesse and Vanden Plas EFi models can be troublesome, and seems to be prone to going out of tune. The electronic control unit fitted to these models can cause problems and is very expensive to replace. The Vitesse does have some disadvantages, with its large front spoiler (on the later cars) and lowered suspension more suited to the race track. It is prone to grounding on speed humps, and is no joke to try and get on the back of a breakdown vehicle. The throttle butterfly bushes in the late twin-plenum chamber Vitesse wear more quickly than in the single-plenum chamber system and are virtually impossible to replace unless you have access to machine-shop facilities.

The 2600 model is a better choice than the 2300, offering more performance and a better ride, as well as more extra features. It is almost as fast as, and more economical than a 3500. The diesel-engined cars are reliable according to owners of these cars, but any savings made on fuel costs are likely to be spent on certain spare parts, which are very expensive, being unique to that particular VM unit. The 'O' Series engine of the 2000 is prone to oil leaks, and the exhaust manifold can suffer from cracking.

The six-cylinder engines have gained a reputation for unreliability – namely head gasket failure or valve gear failure, possibly leading to catastrophic engine blow up. These engines have to be carefully and properly maintained. I have known a number of people who had no problems with their cars, but they maintained the engines meticulously, and changed the oil and filter regularly at least every 6,000 miles (not at the 12,000 mile service interval recommended by the manufacturer on Series II cars), thus preventing sludge building up in the oil ways. On paper, the six-cylinder engine is a very good design with no obvious fault. It is believed that the unnecessary use of jointing compounds when reassembling an engine during a top overhaul can lead to problems – this can choke the lubrication passages through the head gasket, which leads to camshaft failure.

The ROSDI systems marketed by Charter Hydraulics of Gloucester are ingenious systems designed to alert the owners of six-cylinder engined SD1s to the impending failure of the camshaft oil supply, when oil pressure drops. In its simplest form the driver is warned to switch off the engine before any damage is done. Another system activates an external oil supply. You can obtain more details from Charter Hydraulics on 01452 862525.

A number of people transplant a V8 engine to their six-cylinder cars. This task is a fairly straightforward one – the tricky bit is sorting out the wiring and transmission ratios.

When looking for a car, the potential Rover SD1 owner should pay particular attention to corrosion, as this can be extremely expensive to rectify. The paintwork was much improved on Series II cars from 1982 onwards, and the later cars tend to suffer far less from rust than the earlier ones (although there are always exceptions to the rule – the worst SD1 I have ever seen was a red, C-registration Vanden Plas EFi which had supposedly been regularly serviced by a Rolls-Royce dealer – it was so rusty that when the tail-gate was opened, it almost fell off!) Areas to examine closely include the bottom of the rear wheel arch where it joins the sill – this is a favourite rust spot, as well as all four outer wings (these are not easy to replace, being spot-welded) and around the wheel arches. Other areas to watch for corrosion are: the doors, especially the rear doors; the tail-gate, particularly the lower corners and the sunroof (where fitted). The front of the bonnet tends to suffer from paint chips, and can be prone to rusting from the inside. The screen pillars are rear quarter pillars and if the drain tubes have become perished, are subject to corrosion. Another problem can be caused by leaks from the heater (water collects, and this leads to rust).

The rear wheel arch is a noted corrosion area.

Living with the Rover SD1

The trailing lower corner of the rear door is often seen festering away with rust.

In common with most other areas of the body that are notorious for corrosion, the problem with the tail-gate is moisture from condensation running between the inner and outer panels and collecting in the dirt at the lowest point, in this case, the folded seam forming the lower edge of the tail gate.

The common problem of water in the boot of the car is usually assumed to be entering the car through a poorly sealed hatch window. However, water can also get into the car through the poorly designed rear light unit joints. Applying sealant to the gaskets of the rear light units will cure this problem.

Mechanically, the 77mm gearbox is generally pretty robust, and the automatic transmissions – Borg Warner and GM – are usually reliable, although the GM ones can suffer from modulator problems. It pays to look out for noisy engines, excessively oily engines or oil leaks. The suspension is another area to watch out for – unevenly worn tyres can point to suspension parts which may be wearing out. The rear self-levelling suspension units can give trouble – as these are expensive to replace, some people fit uprated gas shockers – so be sensitive to really poor ride quality on the test run. If the rear axle is noisy, give the car a wide berth, as this is costly to replace. The power-assisted steering rack should be checked for oil leaks.

It is worth checking all electrical items to see that these work – that all electric windows go up and down, that air-conditioning and electric sunroof (if fitted) work, and that central-locking, trip computer (these are frequently inconsistent), electric door mirrors, cruise control, instruments, heater and other items are all fully functioning. Electrical items such as the central-locking system are expensive to replace. Any heater fault means the whole dash has to be removed, which is a lengthy job, and can take a couple of days to carry out. Series II cars have plastic-cased front headlamps which, because they sit slightly further forward, are more vulnerable to parking damage than the earlier metal-cased units. Watch out for faulty distributors on Series I 3500s – this all-electronic unit can burn out and cause total ignition failure.

Pay close attention to trim items, as these can be difficult or expensive to replace (although SD1 parts are starting to be remanufactured again). Bumpers in particular are prone to damage – Series I car bumpers frequently have dents, while Series II car bumpers are very fragile and are expensive to replace. The plastic spoiler on the front of Series II cars is prone to damage, particularly on the Vitesse. Check interiors for tears or damage to seats and headlining, damaged door panels, broken parcel shelf and other such items.

How much should you pay for one of these cars? This is a difficult question to answer, as every car should be examined on its merits. A Rover SD1 offers good value for money as a large comfortable car, and is well-equipped. It is much better value, in my opinion, and far cheaper than small, new, modern cars. In a few years' time, very early cars in excellent condition may be worth more than later models. An immaculate car, with a full service history, is likely to fetch a premium price. The late twin-plenum Vitesse models are also likely to fetch more than earlier Vitesses. You can buy a really rough example for a few hundred pounds (but would you really want to?). *Classic Cars* price guide included these figures for SD1 cars:

condition	1	2	3
3500 1976–84	£1,750	£1,000	£500
Vitesse 1982–86	£4,000	£2,400	£1,200
VDP 1982–86	£2,000	£1,150	£600

Note: 1 = excellent all-round condition (concours winners may fetch more)

2 = basically sound but may need cosmetic attention

3 = restoration projects, will be complete and running but will require major work

Living with the Rover SD1

Rover SD1 Club national rallies attract a large cross-section of cars from all variants of the model and in all conditions, but particularly those maintained to a very high standard.

A sea of Rover SD1s parked on the outer circuit banking in the historic setting of Brooklands.

Living with the Rover SD1

People who would like to buy a Rover SD1, but who have never owned one of these cars before, need to know that these cars are costly to run – insurance, fuel and spare parts costs are not cheap (this is one reason why broken-down Rovers often seem to be dumped for months on end – their owners can't afford the high repair costs). The SD1 is not a car to run if you are hard up. Major items like exhaust systems and sets of tyres will cost the owner several hundred pounds – you can either have a large bank balance *or* a Rover SD1, but not both at the same time. Having said that, Rover SD1 parts are considerably cheaper than those of their foreign competitors, particularly German and Swedish ones.

The Rover SD1 is definitely a classic car, and has been increasingly featured in the pages of classic car magazines since the early 1990s. There are two owners' clubs to join – the Rover SD1 Club and the Rover Sports Register. The Rover SD1 Club has well over 1,000 members, and has grown rapidly in the last few years. Its magazine contains useful advertisements from spare parts suppliers. The Rover Sports Register caters for all Rover models. The addresses to contact with an SAE (please check if these are current in classic car magazines) are:

The Rover SD1 Club
PO Box 255
Woking
Surrey
GU21 1GJ

The Rover SD1 Club is frequently represented at classic car shows.

Living with the Rover SD1

A line-up of Rover SD1s at a rally.

Rover Sports Register
8 Hilary Close
Great Boughton
Chester
CH3 5QP

As far as later SD1 cars are concerned, owners who are interested in tracing the histories of their vehicles from British Motor Industry Heritage Trust are unfortunately going to be disappointed. Although they have information about early SD1 cars, all the records of the Cowley-built SD1 cars have been lost, according to BMIHT archivist Anders Clausager.

The SD1 cars (particularly the police cars) were commonly seen in films and TV programmes during their time in production. The view of the general public and *Top Gear* presenters that all Rover SD1 owners must be masochists gained a whole new meaning in the BBC TV series *Blott On The Landscape*, where an MP played by actor George Cole was seen driving a metallic Oporto Red Vanden Plas model, and indulging in flagellation!

Appendix I
SD1 Chassis and Engine Numbers

SD1 chassis numbers

From 1 (a 3500) in 1975.
First 2300: 21477 (1977).
First 2600: 27313 (1977).

Chassis number series, Solihull

1 to 115858 (1979).
120001 (1979) to 137291 (1980).
140001 (1980) to 151137 (end of 1980 model year).
155001 (1980) to 192668 (end of 1981 model year).
200001 (1981) to 215736 (1982, last Solihull car).

Face-lift possibly from 208343 (Jan 1982, a 2600S, believed also first car in calendar year 1982).

First 2000, 208346.
Starting numbers for other derivatives not known.

Cowley-built cars:
250001 (1981–2) to 345831 (a Vitesse; 1986).
First 2600SE: 261625 (?), first Vitesse: 265947 (?).
It is not known whether there are any gaps in the number series for Cowley cars.
All Cowley cars believed to be face-lift model. All cars share one series of numbers.

Total production

Solihull   197,690.
Cowley   95,831.

Total   293,521 (not including any CKD)

(Source: British Motor Industry Heritage Trust)

Appendix I: SD1 Chassis and Engine Numbers

DECODING CHASSIS NUMBER PREFIX

1. 1976–October 1980 models:

First letter,   R = Marque = Rover
Second letter, R = Range = SD1
Third letter,   W = body type = 5-door hatchback saloon
Fourth letter, engine type:
        K = 2300
        M = 2600
        V = 3500 V8.
Fifth letter, model:
        A = 2300
        U = 2600
        F = 3500.
Sixth character is a number, for steering and transmission:
        1 = RHD, 4-speed manual
        2 = LHD, 4-speed manual
        3 = RHD, automatic
        4 = LHD, automatic
        7 = RHD, 5-speed manual
        8 = LHD, 5-speed manual.
Seventh letter indicates model year or major model change; it is always A on 1976–80 models.
Eighth letter indicates manufacturing plant; it is always A for Solihull on 1976–80 models.

2. October 1980 and later models:

VIN code may start with the 3-letter world manufacturer identifier code. This is probably SAR on 1981 models which have 'BL Cars Ltd' VIN plates, and SAX on 1982 and later models which have 'Austin Rover Group Ltd' VIN plates. SAR code also on 1982 models and Vitesse.

Then follows the VIN code prefix proper:
First two letters are always RR – they decode as above.
The third letter is now the model code (previously fifth letter) and decodes as follows:
        A = 2000/2300
        H = 2300S/2400SD/2600S
        F = 2600SE/3500SE
        M = 3500 Vanden Plas, incl. EFi (2600VDP?)
        E = 3500 Vitesse.
Fourth letter is now the W for the body type (compare above).
Fifth letter is for engine type: K = 2300
                          M = 2600
                          V = 3500 V8
                          B = 2000
                          E = 2400 diesel
                          Z = 3500 Vitesse
                          and VDP EFi models.

Appendix I: SD1 Chassis and Engine Numbers

Sixth character is the steering/transmission number, decodes as above.
Seventh letter indicates model year or major model change; it is B on 1981 models (Oct 80–Dec 81, VIN 155001 upwards) and C on Mark II models from Dec 81–1986, VIN 200001 upwards.
Eighth letter indicates manufacturing plant; A on Solihull-built vehicles, to VIN 215736, M on Cowley-built vehicles, from 250001 to 345831.

(Source: British Motor Industry Heritage Trust)

ENGINE NUMBER PREFIXES

The three 'proper' Rover engines have prefixes of a two-digit number followed by a letter. The letter is either A (on all 3500 V8 engines) or C (on 2300/2600 six-cylinder engines).

The following prefixes are recorded:
1. 2300/2600 six cylinder: 10C, 2300 manual
                                11C, 2300 automatic
                                12C, 2600 manual
                                13C, 2600 automatic.
2. 3500V8:
       10A manual gearbox (later 3500SE engine) 9.35:1 CR
       11A auto gearbox (later 3500SE engine) 9.35:1 CR
       12A North American spec, manual gearbox
       13A North American spec, auto gearbox
       14A Swedish spec, auto gearbox
       15A Australian spec, auto gearbox
       16A Japanese spec, auto gearbox
       17A V8-S engine, manual, air con, 9.35.1 CR (later VDP engine)
       18A V8-S engine, auto, air con, 9.35.1 CR (later VDP engine)
       19A Australian spec, manual gearbox
       20A 3500SE with efi for Australia, 8.13:1 CR, manual
       21A 3500SE with efi for Australia, 8.13:1 CR, automatic.

The following codes 23A–28A inclusive are all 8.13:1 CR:
    23A 3500 SE/VDP, air con, manual
    24A 3500 SE/VDP, air con, automatic
    25A 3500 SE/VDP, manual
    26A 3500 SE/VDP, automatic
    27A 3500 SE, air con, manual, for certain hot climates
    28A 3500 SE, air con, automatic, for certain hot climates

The following codes 30A–33A inclusive are all 9.75:1 CR and efi:
    30A Vitesse, manual
    31A Vitesse, automatic
    32A Vitesse, air con, manual

Appendix I: SD1 Chassis and Engine Numbers

33A Vitesse, air con, automatic
(32A and 33A probably fitted to Vanden Plas EFi model)

The 2000 0-series engine has a BMC-type prefix starting with 20V. 20 indicates capacity (2000cc) and V means vertical, which indicates that this is an in-line engine for a rear-wheel drive car. The prefix 20V is followed by a sequence of numbers and letters indicating the exact engine specification, as follows:
    20V/973/AHH = manual gearbox
    20V/974/AHH = Borg Warner auto gearbox
    20V/A18/AHH = GM auto gearbox.

The diesel engine numbers and prefixes are not described in the parts list, but it would appear that the diesel engine has an engine number plate clearly marked 'VM'.

(Source: British Motor Industry Heritage Trust)

Appendix II
SD1 Colours and Trim

Colour	Years	Trim
White colours		
Arum White	83–85	Brushwood, Claret, Osprey
Ermine	82	Bayleaf, Caviar, Oatmeal, Prussian Blue
Pendelican	76–81	Amontillado, Bayleaf, Caviar, Nutmeg, Oatmeal, Prussian Blue
White Diamond	85–86	Bitter Chocolate, Claret
Silver colours		
Argent Silver*	80–81	Bayleaf, Caviar, Oatmeal, Prussian Blue
Platinum*	76–79	Caviar, Nutmeg
Silver Leaf*	82–86	Bitter Chocolate, Brushwood, Claret, Flint, Osprey
Beige colours		
Aran Beige	80–81	Bayleaf, Caviar, Oatmeal
Champagne Beige	82, 85–86	Bitter Chocolate, Caviar
Gold colours		
Cashmere Gold*	82–86	Bitter Chocolate, Brushwood, Caviar, Oatmeal, Sandpiper
Midas*	76–79	Caviar, Nutmeg
Pharaoh Gold*	80–81	Caviar, Oatmeal
Yellow colours		
Barley Yellow	80–81	Caviar, Oatmeal
Turmeric	76–79	Caviar, Coriander
Green colours		
Avocado	77–79	Coriander, Nutmeg
Opaline Green*	82–85	Bayleaf, Brushwood, Caviar, Osprey
Poseidon*	78–79	Coriander, Nutmeg
Silk Green*	85–86	Bitter Chocolate, Flint
Sylvan Green*	80–81	Bayleaf, Caviar, Oatmeal
Triton*	79–81	Coriander, Nutmeg

Appendix II: SD1 Colours and Trim

Red colours
Bordeaux Red*	80–81	Caviar, Oatmeal, Prussian
Carnelian Red	80–81	Caviar, Oatmeal, Prussian
Monza Red	82–84	Brushwood, Oatmeal, Osprey, Sandpiper
Oporto Red*	82–86	Bitter Chocolate, Brushwood, Caramel, Claret, Oatmeal, Osprey, Sandpiper
Richelieu	76–79	Caviar, Coriander
Targa Red	85–86	Caramel, Flint, Osprey

Blue colours
Atlantis	77–79	Caviar, Coriander
Azure Blue*	85–86	Flint, Ice Blue, Prussian Blue
Caribbean*	76–77	Amontillado, Caviar, Coriander
Cavalry Blue	80–81	Caviar, Oatmeal, Prussian Blue
Eclipse Blue	83–85	Bounty Blue, Osprey
Moonraker Blue*	82–86	Bounty Blue, Flint, Ice Blue, Oatmeal, Osprey, Prussian Blue
Nightwatch Blue	86	Flint, Ice Blue, Prussian Blue
Persian Aqua*	78–81	Caviar, Coriander, Oatmeal, Prussian Blue
Zircon Blue*	82–85	Bounty Blue, Brushwood, Oatmeal, Osprey, Prussian Blue

Brown colours
Brazilia	76–79	Amontillado, Coriander, Nutmeg
Clove Brown	83–86	Brushwood, Caramel, Sandpiper

Black colours
Black	79, 82–86	Caramel, Caviar, Claret, Coriander, Nutmeg, Oatmeal, Osprey
Maraschino	80–81	Bayleaf, Caviar, Oatmeal, Prussian Blue

* denotes metallic colour

(Source: Leyland Cars and Austin Rover colour cards)

Further Reading

Adeney, Martin (1988) *The Motor Makers – the turbulent history of Britain's car industry*, Collins.

Alder, Trevor (compiler) (1993) *Rover 3500 SD1 – the early years*, Transport Source Books.

Alder, Trevor (compiler) (1996) *Rover SD1 in motor racing*, Transport Source Books.

Alder, Trevor (compiler) (undated) *Rover SD1 2000, 2300, 2600 & 2400SD Turbo Diesel*, Transport Source Books.

Beehl, Nathan (1996) *Ferrari Daytona*, Haynes.

Clarke, R. M. (compiler) (undated) *Rover 3500 & Vitesse 1976–1986*, Brooklands Books.

Daniels, Jeff (1980) *British Leyland – the truth about the cars*, Osprey.

Edwardes, Michael (1983) *Back From the Brink*, Collins.

Georgano, Nick (ed.) *Britain's Motor Industry – the first hundred years*, GT Foulis & Co.

Hardcastle, David (1990) *The Rover V8 Engine*, Haynes.

Hough, Richard and Frostick, Michael (1966) *Rover Memories*, John Dickens & Co.

King, Peter (1989) *The Motor Men – pioneers of the British car industry*, Quiller Press.

Long, Brian (1993) *Standard – the illustrated history*, Veloce.

Mowat-Brown, George (1992) *The Rover*, Shire Publications.

Pagnamenta, Peter and Overy, Richard (1984) *All Our Working Lives*, BBC Books.

Robson, Graham (1988) *The Rover Story*, Patrick Stephens.

Robson, Graham (1991) *Triumph TRs*, The Crowood Press.

Robson, Graham (1995) *Triumph 2000 and 2.5PI*, The Crowood Press.

Robson, Graham and Langworth, Richard (1988) *Triumph Cars – the complete story*, Motor Racing Publications.

Taylor, James (1996) *Classic Rovers 1945–1986*, Motor Racing Publications.

Taylor, James (1993) *Rover P6 1963–1977*, Motor Racing Publications.

Taylor, James (1991) *Rover SD1 1976–1986: owner's & buyer's guide*, Yesteryear Books.

Wood, Jonathan (1988) *Wheels of Misfortune – the rise and fall of the British Motor Industry*, Sidgwick & Jackson.

Young, Daniel (compiler) (1993) *Advertising Rover, Volume II 1904–1984*, Yesteryear Books.

Index

AA (Automobile Association) gold medal 12, 55
ABS 33
AEC 83
Alfa Romeo 107
Allam, Jeff 128–9, 140
Allinson, Peter 154
Archer, Dan 154
Arikkala, Pentti 132
Arthur, Rob 133–4, 136
Aston Martin 38, 58, 119
Atkin, Graham 106
Audi 46, 54, 56, 65, 67, 69, 76, 93, 106, 119, 125
Austin models:
　Allegro 29, 68, 81, 165
　A30 30
　Maestro 30–1
　Maxi 25, 81–2
　Metro 30
　Metro 6RA rally car 135
　Mini 71, 161
　Montego 106, 157
　Princess 69, 81, 103, 156–7
Austin Rover 100–1, 106–7, 111–13, 118–19, 121–2, 124–7, 130, 188
　Austin Rover Rallysprint 131–3
Australian SD1 see Rover SD1 models
Autocar 43, 45, 48–51, 62–3, 65–7, 69, 71–4, 76, 84, 86–7, 93, 98, 103, 105, 109–10, 112, 115, 119, 130, 162, 166
Autosport 52, 68, 120, 130, 133, 138, 140, 167
Axe, Roy 6, 165

Bacchus, John 6, 155–7
Bache, David 10, 21, 23, 25–32, 47, 82, 86, 171–2
Bahco 129
Barber, John 23–4, 81–2
Barbet, Dennis 6, 106
Bashford, Gordon 23–4, 30–1
Beans Industries 41
Beech, Ian 29
Bell, Roger 68
Bentley 10
Beveridge, Ian 6, 139
Bianchi Rally 133, 139
Blomqvist, Stig 132
Blott On The Landscape 182
BMC 157
　BMC 1800 33
BMW 51–2, 54, 56, 70, 84, 93, 119, 128–30, 140–1, 167, 169
Borg-Warner automatic transmission 38, 44, 51, 98, 103, 179, 186
Bolster, John 52, 68
Bosch 107–8
Brabham, Jack 20
Brands Hatch 128, 140
Bristol 51
Broadspeed 106
British Grand Prix supporter 140
British Insurance Association's Research Centre 34
British Leyland (BL) 7–8, 12–13, 16, 21–3, 30, 32, 37, 41, 55–9, 69–72, 79–82, 99–100, 105, 158, 161, 165, 167–9, 189
　P76 model 22, 156
British Leyland Mirror 21
British Motor Industry Heritage Trust (BMIHT) 6, 42, 87, 124, 158–9, 161, 170, 182–3, 185–6
British Rail 81
Brooklands 180
Brown, Peter 133
BSCC (British Saloon Car Championship) 118–19, 127–30, 137, 139, 140
Buick see GM

Car 52, 54, 56, 73, 77, 84, 86, 111, 119, 166–7
Car of the Year Award 54, 71, 87
Car of the Year, Today 27–8
Care le Gant 162
Charles, Prince 159
Charter Hydraulics 177
Chrysler Alpine 26
Cibie 35
Circuit of Ireland Rally 133, 139
Citroen 49, 65, 76, 101, 106, 110
Clarke, David 113
Clarkson, Jeremy 7
Classic Cars 166, 175, 179
Clausager, Anders 6, 182
Cole, George 182
Coleman Milne 161
Collins, Peter 111
Cook, Mike 6, 155
Country Life 56–7
Coventry Climax 38

Cumbria Rally 133

Daimler 83
Davenport, John 6, 13, 113, 116–18, 124, 127–33, 135, 137, 139, 144–5, 167
Day, Graham 167
De Dion rear suspension system 10, 16, 22, 33
Denovo tyre see Dunlop
Desai, Vice President 158
Donington Park 129, 132–3, 137, 140
Don Safety Trophy 54
Dove Group 161
Dunlop
　Denovo run-flat tyre 12, 34, 44, 46–7, 54–5, 64, 93, 95

East Hamphire Post 52–4
Edis, Alan 6, 23, 159–61
Edward Grace International 129
Edward, Prince 118
Edwardes, Sir Michael 58, 81, 159, 161, 189
Eley, David 6, 23, 36–7
Empire Trophy 128
Epynt Stages Rally 133
'E' series engine 156
Esgair Daffydd Rallysprint 131, 133, 135
Estate car see Rover SD1 models
Estoril race 140
ETC (European Touring Cars) events 128–30, 137–8, 140

Facel Vega 30

190

Index

Ferguson tractor 16
Ferrari 25, 27–9, 171–5
  Daytona 7, 26–7, 171–5
Ford, Henry 17
Ford models
  Capri 128, 140
  Cortina 101
  Escort 131, 139
  Fiesta 54
  Granada 7, 45–6, 56, 69–70, 87, 98, 101, 104–6, 145, 169–70
  Orion 7
  Sierra 152
Ford Motor Company 6, 17, 23, 56, 161, 169
Fowden, Bob 133
French, Richard 6, 145, 152
French Saloon Car Championship 140

gearbox, 77mm 15, 32, 36–7
General Motors (GM) 11, 18–19, 180
  models
    Buick 7, 10–11, 17–20
    Chevrolet 17
    Pontiac 18
    Oldsmobile 18
GKN 37, 135
Grant, Peter 80
Greenslade, Rex 127–8

Harvey, Tim 129, 140
Haynes Motor Museum 161
Hepworth and Grandage 40
Herbert, Jeff 84
Heslop, Nigel 6, 70, 169
Hill, Graham 12
Hindustan Ambassador 157–8
Honda 8, 94, 139, 165–6, 169
Hopkins, Keith 56
Horrrocks, Ray 80, 127
Hot Car 164
Hunsruck Rally 133

Jaguar 10, 12, 15, 22, 36, 46, 49, 51, 66, 84, 100, 125–8, 141, 151

Jaguar Rover Triumph 84, 100, 127, 155
Jamara race 140
Janspeed 6, 151–2, 162–4
  Janspeed police SD1 151–2
Jensen 27–8
Jones, Alan 140

Keikhaefer, Carl 18
King, Spencer ('Spen') 6, 10, 12, 20, 23, 25–6, 30–1, 33, 35, 82, 103, 113, 156, 168
Kitcars and Specials 175

Lancia 51, 54, 65
Lada 82
Leach, Dennis 140
Le Mans 12–13, 173
Lewis, Edmund 9
Lewis, Graham 6, 95, 117
Lewis, Mike 6, 21, 23, 28–31, 33–4, 41, 168
Leyland Australia 155
Leyland South Africa 156
'Lisa 10' (South African SD1) 156–8
Loasby, Mike 6, 23, 38–9, 41
Longbridge pplant (Austin) 81
Lotus 121–2
Lovett, Peter 140
Lucas 35, 79, 114, 116, 153, 155

MacPherson strut 33, 35, 44, 61, 102, 108, 114, 116
Malkin, Colin 131
Mander, Phil 6, 23, 34
Mansell, Nigel 132–3
Manx International Rally 133
Martin-Hurst, William 10, 12, 18–19, 20
Marvin, Rex 6, 23, 32, 34
Maserati 27–8
McRae, Jimmy 132
Mercedes 7, 22, 27–8, 34, 46, 49, 51–2, 54, 56, 84, 87, 93, 110, 119, 167

Mercury Marine 19
Metge, Rene 128, 140
Metropolitan Police *see* Police
Michelotti 82
Middle East rallies 131
'Midlanders of the Year' award 31
Ministerial cars 10, 151
MIRA 35
Modern Motor 155
Monza race 140
Moore, Roger 58
Morris 12, 81, 103
  models
    Marina 81–2
    Oxford 158
Motor 46, 48–9, 51–2, 60, 68–9, 74, 87, 94, 99, 104–5, 111–12, 120–1, 124–6
Motor Shows 20, 77, 94, 107, 113
Motor Sport 52, 69–70
Motorsport, Rover SD1 127–40
Murrough, Peter 127–8
Musgrove, Harold 99, 113, 125, 139, 167

Nogaro race 140
North American SD1 *see* Rover SD1 models
Nuffield, Lord 81

Opel 84, 87, 93, 119, 139
'O' series engine 101, 103–4, 106, 186
Oulton Park 140

Panhard rod 33
Paris–Dakar Rally 131
Percy, Win 129, 138, 140
Performance Car 167
Perkins 107
Peugeot 33, 46, 51, 56, 65, 67, 110
Phillips, Captain Mark 118
Pininfarina 82
Pirelli 35, 116
Police 12, 141–52
  Cheshire 144
  Cleveland 143

Metropolitan 145–52
Northamptonshire 141, 143
Surrey 143
Thames Valley 141, 144–5
West Mercia 143
Wiltshire 141
Yorkshire 36, 141–3
Police cars 12, 141–52
Police Review 141
Pond, Tony 6, 117, 131–40
Porsche 54
Pressed Steel Fisher 23, 32
Price, David 128
  David Price Racing 127–8, 140
P76 *see* British Leyland models
Purkis, Geoff 6, 27–9

RAC 6
RAC enquiry 129–30, 140
RAC Tourist Trophy 12, 128, 140
Reignier, Peter 139
Renault 25, 46, 54, 56, 67, 82, 87, 99, 101, 106, 132
Repco 20
Rivers, Jean 79
Road and Track 154
Robin Hood Daytona 174–5
Robin Hood Engineering 175
Robinson, Derek ('Red Robbo') 81
Robson, Graham 6, 16, 18, 24, 158–9, 172, 189
Rolls-Royce 10, 51–2
  Silver Shadow 52
ROSDI systems 177
Rouse, Andy 128, 131, 140
Rover
  safety bicycle 9
  V8 engine 17–20
Rover-Honda XX project 94, 165
Rover models
  Ten 9
  Light Six Sportsman's Coupé ('Blue train Rover') 9

191

Index

P1 31
P2 9–10
P3 9, 31
P4 ('Auntie') 9–11, 16, 22, 30–1
Scarab 31
P5 and P5B 10–12, 19–20, 22, 30–3, 151, 166
P6 10, 12–16, 20–1, 30–5, 43, 45–6, 55–6, 58, 60, 70, 103, 141, 145, 149–50, 166, 168–9
P6BS 30–1
BRM *see* Rover racing
P8 22, 24, 31, 33
P10 16, 21–2, 24
RT1 21, 24
SD1 Australian 155–6
SD1 CKD 156–7, 170, 183
SD1 estate car 159–61
SD1 Indian *see* Standard 2000
SD1 North American 153–5
SD1 Six cylinder cars (2300 and 2600) 59–70, 89–100, 141–52, 156, 166, 168, 170, 176–7, 184–5
SD1 South African 156–8
SD1 2000 153, 156, 101–6, 166, 170, 176
SD1 2400SD Turbodiesel 107–12, 166, 170, 176
SD1 3500 43–8, 89–100, 141–52, 166, 168, 170, 176–7, 179, 184–5
SD1 Vanden Plas 92–100, 166, 170
SD1 Vanden Plas EFi 124–6, 166, 170, 176–7, 186
SD1 V8-S 92, 83–8, 153, 166, 170, 185
SD1 Vitesse 113–26, 128–9, 131, 135, 138, 140, 149–52, 156, 166–8, 170–9
SD2 *see* Triumph models
800 Series 135, 139–40,
152, 165–7
Land Rover 20, 30–1
Range Rover 15, 20, 30
Rover racing 12, 20, 113, 115–24, 127–40
Rover BRM 12, 20
Rover rallying 12, 127–40
Rover SD1 Club 6, 180–1
Rover SD1 plants
  Blackheath, Cape Town 156
  Castle Bromwich 72, 77
  Coventry 37
  Cowley 72, 81, 89, 182–3, 185
  industrial problems 41, 71, 77–82
  Pengum 37
  Solihull 27, 52, 71–82, 183, 185
  Solihull paint plant 71–2, 75, 78–81
  Swindon (body panels) 80
Rover Sports Register 6, 181–2
(The) Rover Story 18, 24, 172, 189
Rover-Triumph 23–4, 84
Royalty Protection cars 149, 151
Russek Manuals Rally 133
Ryder report 57–8

Saab 119, 125
SCCA races 13
Scottish International Rally 133
Scottish Rally Championship 133
SD2 *see* Triumph models
Six-cylinder engine development 37–41
Sked, Gordon 165–7
Skoda 7
SSMC Engineering 161
Seal, Roland 31
Shawcross Enquiry 129–30, 140
Shell Oils Open Rally Championship 133–5, 138
Silverstone 199, 127, 129,
137, 140
Snowden, Mark 139
Soper, Steve 128–9, 140
South African SD1 *see* Rover SD1 models
Spa 24 Hours race 137–8, 140
Specialist Division Number One 7, 24
Stabilimenti Meccanici VM 107, 176, 186
Standard 9, 15–16
  models
  Vanguard 15–16, 158
  2000 (Indian SD1) 15–16, 157–9
Standard Motor Products (of Madras) 157
Standard-Triumph 15
Starley, John 9
Stewart, Jackie 12
Stewart, Richard 175
Stokes, Lord Donald 12, 24, 57, 82
Style Autto award 55
Sullivan, Danny 132
Sytner, Frank 129

Talbot 106
Tales of the Unexpected 166
Taylor, Trevor 117
Thomas, Hywel 133
Thruxton 140
Top Gear 7, 182
Tow Car of the Year award 54
Timken 36
Transport, Department of 6, 144–5
Triplex Ten-Twenty safety windscreen 47, 54, 89
Triumph models:
  Acclaim 81
  Dolomite 13, 15, 30, 38, 62
  Herald 13
  Puma 24, 82
  SD2 82, 169
  Spitfire 13
  Stag 15, 30
  TR2 13
  TR7 15, 36, 72, 82
  TR8 13, 15, 72, 127
2000/2500 13–16, 21, 24, 33–4, 40, 60, 70, 168–9
  (Gloria) Vitesse (1930s) 12–13
  Vitesse (1900s) 13
  Triumph 2000 and 2.5PI 16, 189
Trossachs Rally 133
(The) Two Ronnies 81
Turley, Joe 19
Turner, Philip 60, 62, 74
TVR 20
TWR (Tom Walkinshaw Racing) 128–40

Ulster Rally 133

Vallelunga race 140
Vanden Plas (company) 92
(for Vanden Plas cars *see* Rover SD1 models)
Vauxhall 33, 125, 169
  Cavalier 110, 169
Vitesse *see* Rover SD1 and Triumph models
Volkswagen 18, 54
Volvo 34, 46, 51, 54, 56, 65, 67, 87, 93, 99, 125, 129, 140, 161

Walkinshaw, Tom 128–9, 131–3, 137–40
Warwick, Derek 132
Watson, John 132
Watts linkage 33, 44, 77, 135
Welsh International Rally 133, 135, 139
Welsh Tarmac Rally Championship 133
What Car? 55, 105, 110, 119, 126
Wilks, Maurice 9, 12
Wilks, Peter 12
Wilks, Spencer 9, 12
Williams 111, 132
Wilsons U-Bix Kingdom Stages Rally 133
WL9 racing camshaft 122, 124
Wolseley 150
Wood and Pickett 161–3
Wood, Ken 133

192